About the Author

Roberto is the pen name for Robert Radulescu, born in Giurgiu Romania in 1971. Currently Roberto is an Irish citizen living in Naas, County Kildare. Roberto works in Social Care in a managerial post and is the father of one. He has a passion for science, history, religion and science fiction, film and literature, from a very young age. His influences come from prominent philosophers, writers, scientists and film directors and producers such as Plato, Dante Alighieri, Immanuel Kant, Jean Paul Sartre, Albert Einstein, Isaac Asimov, Jules Verne, Karl May, Erik Van Daniken, Carl Sagan, George Lucas, Ridley Scott just to name a few. As well music is a big part of his life as he is a declared fan of hard rock and heavy metal.

Dedication

For all family and friends, thank you for your support

Roberto

Swallowed...Into The Pit

A CIP catalogue record for this title is available from the British Library.

ISBN 978 1 78455 547 4 (Paperback)
ISBN 978 1 78455 549 8 (Hardback)

www.austinmacauley.com

First Published (2015)
Austin Macauley Publishers Ltd.
25 Canada Square
Canary Wharf
London
E14 5LQ

Printed and boun

Acknowledgments

I would like to thank to all those who supported me in writing this book, which include my son Sasha and lifetime partner Silvana first of all. Of course I say thank you to all those who not knowingly helped me shape my vision and ideas concerning life generally but also on matters such as philosophy, religion, science etc. By the way these are many from writers and thinkers of ancient times to people that are more or less contemporary, but somehow had a major impact on this world and implicitly on my own life. To perhaps mention a few names I'd start with Plato, Kant, Dostoevsky, Sartre, Sagan, Lucas, von Däniken and many more. As well I have been blessed to have been raised in a family most avid of knowledge and especially my granddad (God rest his soul), and my father who put many books into my hands from an early stage in my life. Finally but not least, I'd like to thank Austin Macaulay Publishers for welcoming me as one of their authors, making this book possible, and most important to acknowledge their high standard of professionalism while offering a very friendly publisher-author relationship. Thank you!

Disclaimer

Any historical event, names of historical figures and locations are being used hereafter in a purely fictional manner and do not claim to have anything to do with real events, people and places. This is a fictional novelistic piece of literature, in the sci-fi genre and should therefore be treated as such…

THE AUTHOR

Factual

In 1968 Swiss author Erich von Däniken published his first book "Chariots of the Gods?" in which he asked a number of questions that ignited theorists from many areas of research, creating an unprecedented controversy, the so called Ancient Astronaut Theory. Although discredited by many, the theory is still as popular today asking valid questions for which no science has found any plausible answers. To me personally the theory represents an alternative to main stream history and archaeology, sciences that would not consider conducting research into these issues from a different stand point. I am however hopeful that one day they will. I base this on what happened to Astronomy not long ago. Essentially astronomers could not even entertain the idea of life elsewhere for the simple fact that until early 90's they had no evidence of exoplanets orbiting other stars and moreover our own solar system was considered to be "a freak of the universe". Once the first exoplanet was discovered, not only could astronomers consider extra-terrestrial life but they are now 100% sure of it with a catalogue of exoplanets getting larger on regular basis and with planets observed to be in the habitable zone of their stars. It perhaps takes a significant discovery for researchers in other areas of science to change their approach on things and it will perhaps happen to main stream archaeology one day too. I am looking forward to that day…

Chapter One

It was hard for Vincent to comprehend what was happening to him when he became conscious after what appeared to have been a long time of deep sleep. He could only see this foreign face of a dark skinned man, who was shouting something in a tongue that Vincent could not make out. Suddenly he sensed another presence nearby but he felt way too stiff to try and make visual contact. Something was the same such as the jungle which he could recognise, but also different since this weird sensation of not belonging here, not in this world, was overwhelming him. That strong feeling raised one essential question, what world was this? And speaking of worlds and belonging, well Vincent had some remarkable and equally dramatic experiences of such, experiences that he could not place in any particular space-time dimension as his confusion was way too deep right now. He swallowed this lump and felt in his throat trying to somehow shun all his doubts and rather try to concentrate on whatever it was that was happening to him now.

'He said that you were delivered by the gods,' this woman's voice said in plain English.

Vincent could not respond but instead he heard himself mumbling something. It was only the word "gods" that hit something in his brain only to make him shiver. Because his head was filled with memories of "gods" if that word could describe anything near to what he experienced.

'How are you feeling,' the same women's voice asked at the same time as the dark skinned man got out of his sight. A cold, wet cloth began to gently wipe his forehead which made him feel better.

'Can I ask who you are? Because I have to tell you, your presence here is absolutely bizarre to say the least,' the woman who Vincent could see now, spoke again.

She was beautiful, he thought, cute face, dark with deep green eyes, black hair left loose on her nicely shaped shoulders. He didn't know what his response should be, not in that split second he took to accustom himself to the beautiful woman, but then his disciplined reaction kicked in.

'I am Colonel Vincent Monroe of the United States Army, Second Regiment Cavalry, under direct orders from President William Howard Taft, ma'am,' he said in complete military fashion.

'I can see that you are in a deep state of confusion,' the woman began with a sense of sympathy in her voice, 'the locals found you at about two miles south from here. I must say that you are very lucky to be alive,' she finished in a more optimistic tone.

Something was terribly wrong Vincent could sense. He had the certainty that he did not belong here whatsoever, but the question was, "here" meant where? And yes, it was the same jungle, but it was in a different time. How long had he been lost in the forest? How long had he spent with the "gods"? Whatever happened to him, he could not explain in ordinary terms, but surely there must be an explanation.

'Who might you be Miss?' Vincent whispered feeling frail in his own thoughts.

'I am Doctor Linda Wilson, and about your president, well I can't even tell when the man passed away but it was a long time ago.'

'The President, is he dead?!' Vincent mumbled in a sheer tone of surprise. He knew that he might have been lost for quite some time, but still he felt it wouldn't be so long that the last president known to him, and the very man who sent him here was long gone now, as the woman claimed.

'Look,' Linda interrupted his thoughts, 'two days from now a helicopter will drop us supplies and I have already arranged for you to be taken to Peru. An army carrier will then transport you to the US. I am limited in terms of medicines, I am here on a research project, you know?' Dr Wilson explained.

'A heli what..., what did you call it?' Vincent asked purely baffled.

'Uh, a transport by air, look Vincent, if this is your name, I have no idea how you got here, and the truth is your clothing style does not belong in this time. I have also noticed some of your other belongings which you had in a bag, now I didn't go through them, but just noticed that they are strange, the fact is I cannot look after you, not properly because I think that you need a lot of help whatever it is that happened to you.' Linda explained with conviction.

'Is my diary in the bag still, because that is one precious thing that I must bring to the President,' Vincent rushed to say, 'well we still have a president don't we?'

'Yes Vincent, we do. But before you reach him I need you to rest and then to be fit for the journey ahead. I have treated most of your wounds, mostly cuts and bruises, some looked older and hard to say how you acquired them, while some were new and infected however you are fine now. Back home they will give you all the medical care that you need, I'll make sure of that.'

Minutes later Vincent and Linda were inside what appeared to be a colourful tent, after he was helped to walk across a few yards. It was at this moment that Vincent noticed a big table with all sorts of tools and apparatus that were foreign to him. They seemed foreign all right, but some still familiar because he was sure that he had encountered that sort of gear, but could not tell when, maybe just in some past visions he had, terrible visions it must be said. Some of these devices looked like rectangular colourful boards, but somehow letters of text and weird pictures which were not inscribed by anyone but rather contained inside the boards, were showing up. The only thing that he could have guessed rightly about these machines was that they were powered by electricity. Vincent could not name them and felt somewhat embarrassed, even afraid, to ask Dr Wilson about them. Besides, his mind was rather concerned with other issues, such as his current state of affairs and his near future. And that being the case he felt that he was being left at the mercy of this female doctor, which bothered him quite a bit. It wasn't often that he felt powerless in front of fate, usually as an officer and a commander he would be in control of pretty much every circumstance. However he had to admit that there were more than a few situations in his life when he had to lie down arms in front of fate. And for that matter something very dramatic must had happened in a past that he could not establish still, but he felt it was recent, in spite of waking up to a completely different world. What this dramatic event was would again be something hard to establish. Surely whatever it was had some disturbing effect on his whole being, often generating powerful emotions associated with horrible visions of places, of beings, all sorts of beings, of machines hard to imagine and now this place. How long had he been lost in the darkness, for now even the President was long dead? Indeed there were so many questions that needed answers, but Vincent knew that many of these answers were somehow buried deep within his own soul and mind. The only dilemma now, was for him to figure out a way to reveal them in the light of consciousness, in a way that they could be explained.

'Doctor Wilson, if I was to tell you that I've suffered things that I cannot understand, I cannot put my thoughts into words in order to explain, neither what happened nor how I feel, would you believe me?' Vincent cried, speaking his mind.

'Oh God, surely whatever has landed you here Colonel is odd, but I can't help you more than treating a few wounds and that sort of thing. However, back in the US they will take care of you properly. I will write a letter explaining the circumstances in which you were found, as well as to request a psychological evaluation right away on your arrival,' Linda explained.

'Thanks Doctor, that's so kind of you, whatever that evaluation of yours means, but thank you,' he uttered in response.

There were so many things he could not understand, since he woke up in this strange world. He felt that the doctor spoke strangely, using words that were not common back in his time. How much more was there about all this, strange tongue, strange electric gear that writes words by itself, air transport called "heli" something and so on? It did not matter much anyway, not now because a sudden tiredness slowly settled back into his bones and mind, having him deviate from reality. He came out of it, on and off during two rather agonising days. He didn't ask any more questions, nor did he engage in long talks with the female doctor, but just kept strictly to a few words about his needs and such-like. The doctor was always busy anyway, running around the locals' village, shouting words in their native tongue, sometimes coming in and out of the tent, picking up things. However not once did she come in without taking an interest in Vincent's wellbeing, and then she would sleep in the same tent late at night. This other woman, a local who spoke no English, was helping him feed, wash and even take a few steps outside the tent. All that was happening until this "thing" plunged from the sky, it thundered and made the trees bend, as it slowly came down to a stop, on the fresh grass only a short distance from the tent. It was a strange looking machine, with something like a big rotating hat at the top. Vincent could see a couple of men coming out of it, who began carrying many boxes, small and large, out of the "flying beasts' belly", that they laid down on the grass. He then saw Dr Wilson speaking to the men, gesturing towards the tent after which she came inside and told him that the flight would be so much easier if he was sedated. He didn't like the idea of that after she thoroughly explained what it meant, but he trusted her, simply because he did not have anybody else to trust, not even himself that he could rely on. He felt sad having to leave the company of the cute female doctor, although they only had a short time together, he felt like he had a bond with her, for the simple fact that she appeared to be reasonable to his peculiar situation. She did say to him in passing that she would be in Washington soon and would come to visit him, but once again felt helpless not knowing if he would ever see her again. The next twelve hours or so, after seeing Dr Wilson for the last time, were rather unbearable for Vincent. He drifted into unconsciousness and came back a few times during the journey to Washington, hardly being able to tell where he was or what was happening to him. He only knew that he was now in a large military aircraft. Occasionally strange men dressed in strange uniforms were addressing him about what he needed and nothing else of relevance happened until he woke up in this clean room on a comfortable bed with fresh linen. It was a little before lunch time, judging by the illumination from the sun, which was flooding through a large window. He felt a bit dizzy, not unpleasantly though, but above all he was fully aware for the first time in God knows how long. He sensed that he was alone in the room however there was this weird sensation of being watched. After all Colonel Monroe was a man formed by military training and discipline, combat situations and long term campaigns, especially if counting his latest adventure. Time and time again life and death situations demanded that his senses be fully attuned, or else he wouldn't be walking and talking anymore. Therefore his sensation of being watched must be correct, but it felt different to many other times, not only because he was confined in a room, but who would want to watch a weakened man wearing a strange robe? Instinctively Vincent made a few steps towards the window, watched from behind the curtain side, trying to make sure he is not seen by anyone on the outside. All he could see though were a couple of nicely trimmed bushes and trees, placed on a large lawn area divided by paved alley ways. He managed to ease down his initial panic, however the sensation of being watched was still lingering deeply in his bones. As it turned out Vincent wasn't at all wrong, because the moment he turned around to face the door, he noticed this strange black globe like object, hanging just above the door frame. Even scarier the object had a round shaped glass lens, cut right in the middle of it, but it also appeared to move, following his own movements around the room. He attempted to run to it but an unnamed powerful sensation kept his feet pinned to the floor. He was terrified by this strange object, not only because he knew now that he was "watched", but rather because he had experienced something similar before. And the paralysing fear was generated by his subconscious memory, essentially ordering his entire being not to attempt to move, or else consequences would be very painful. But after a few minutes of waiting nothing happened, so with his breath almost smothered by his overwhelming fear, Vincent decided to walk to the door and see what lay on the other side. He felt a bit childish in questioning his mind about the unknown on the other side of the door, but his fears were well embedded in his very soul, likely for good reasons. As he made a few weak steps towards the door, this opened in front of him only to reveal a middle aged woman, dressed casually, smiling delicately, but confidently all the same. She scanned him from top to toes in one split second, then closed the door behind her, while indicating he should sit down on the bed.

'Please, Mister Monroe I believe, make yourself comfortable,' she said while taking a seat on a nicely crafted wooden chair set right inside the door. 'We need to talk, so please sit down.'

She sounded very assertive, even a little intimidating for Vincent but he followed her instructions and sat on the bedside.

'Welcome to Saint Elizabeth's first of all, I am Doctor Kate Palmer from D.C. Department of Mental Health and I don't...,' she attempted to say before being interrupted.

'I don't recognise this place, is this in Washington?' Vincent asked his voice innocent.

'Yes, it is in Washington, have you by chance been here, in this facility before? '

'No I don't think I have, but please tell me what that is above the door,' he enquired keeping his innocent tone.

'Which..., huh the camera you mean? You see, we only monitor new comers just for safety reasons, but hey we are not invading your privacy, besides the bathroom is camera free, so not to worry,' Kate explained sounding reassuring.

'Monitor, meaning what, I mean is that thing watching me?' Vincent insisted uncertain of what the doctor was saying.

'In effect, yes it does, but only a few minutes at a time, so that we know you to be safe..,' Kate further reassured.

'What is this place, cause I don't feel very safe, is it like a treatment centre?'

'Yes that's right, for people with trouble of the mind, shall we say?'

'But I don't feel any trouble, it's just that I am rather a little confused, that's all.'

'Well confusion is a state of the mind therefore you are in the right place. Let me tell you in return that I am confused too, as to how you got here, huh what am I saying? Look Mister Monroe you were flown in from Peru, landed in Washington at twenty two hundred hours last night. That was after you had been found by a local tribe just inside the Brazilian border, about three days ago, and luckily there was an American doctor there who is running a research project with the locals via a government funded programme. Anyway, my question for you is how did you get to be in the heart of the Amazonian jungle?'

'I remember Doctor Wilson, nice lady, but she told me I would meet the President on my return to Washington, so if I can see him, I have a very important report to give him, and yeah I know it's a different President nowadays, but nevertheless my testimony is equally precious. Uh, by the way can you tell me what year this is, I meant to ask that other doctor but I felt a little embarrassed, so to say,' Vincent explained, his innocence very disarming.

Doctor Palmers looked at him in total bewilderment just about realising the extent of the man's delusion, not that it would be her first encounter of this kind however the circumstances in which this man landed in her backyard were quite unheard of. She took a sigh, hardly perceptible, as she didn't want to betray her astonishment and then tried to devise some sort of response that would not add to Mr. Monroe's mystification, but neither it would dismiss his question.

'Okay, Mister Monroe...,'

'You can call me Vincent, just to make it more amicable, so you were saying,' he abruptly intervened, his voice revealing some kind of uneasiness.

For Dr Palmers this was a good indication that this man, Vincent if that was his real name, wanted to make a connection with her by bringing down the formal barriers. She had to accommodate that because even in the short time she has spent with him, she had to admit that her fascination with his case was running wild. She had to also acknowledge that she had begun to feel a little weird in his company, some kind of anxiety that she never felt in the presence of any other patient all through her career. This guy was just different in all respects concerned. That he wasn't well, was just one thing, a minor thing, as for everything else "Vince" was just one huge mystery for which she needed tact, patience and perhaps loads of sympathy, positive influence at all costs, in order to decipher it. She felt that it was just the start of a long hard road ahead, but at the end of which a massive reward was waiting. She made an imperceptible effort to regulate her breath, put on a smile and then began in a soft voice.

'Okay Vincent, you can call me Kate in return, but look, I need to establish a few facts before I will be able to help out with your situation, which is rather strange you'd have to say. So just to recap, you were found in the Amazonian jungle a few days ago by a local tribe. Luckily Dr Wilson was there so she helped out to send you back home, because Vincent you are an American, is that correct?'

'So far it's all good, ma'am!'

'Any chance you'd remember how you got there, cause for us here, this is the most confusing part if you get my meaning.'

'Perfectly clear, I was sent there by President Taft with a mission, about which mission I am not authorised to speak, not to you but only to report back to the President.'

'Is that Howard Taft, the President, because oh boy..., he died, a very long time ago, you know? What year did you go on your mission Vincent?' Dr Palmer asked, her voice a little irritated.

'Nineteen twelve, ma'am, I do realise that it's much later now, judging by all these machines that I saw and after speaking to Dr Wilson, but really what year is this, cause I have a feeling that I was supposed to be dead,' Vincent replied assertively.

'Yeah, you're right, perhaps long dead now Vincent, cause we are in twenty thirteen, but as a matter of curiosity what age were you when you set course for the Amazons?'

'Forty two, I was forty two back then, it does sound odd, but what can I say? I am here speaking the truth, which terrifies me too. There again nothing is so peculiar, oh no ma'am, not at all, not after what I've been through.' Vincent said sounding convincing.

Dr Kate Palmer took a few moments trying to analyse his words, body language and facial expression but all appeared that of someone who spoke the truth, or at least he appeared to be convinced of what he was saying. This sort of thing she had encountered before, many times as it was common with schizophrenics, but still she felt that she wasn't in front of a classic example of that. And more importantly the circumstances in which Vincent was found were out of the ordinary. Was it possible that this man had been part of a more recent expedition in South America that went wrong; God knows perhaps he got lost for long enough that his mind took a turn for the worst? This was quite plausible, she thought, but she decided not to suggest that to Vincent, as she felt that in order to establish his level of delusion, it was paramount to hear his story. She posted a frail smile, once more, to suggest that she believed him and then nodded.

'Vincent, before talking some more, would you consider having breakfast? I will talk to my boss about your case, not to worry, it's just a formality and then I will meet you in his office in about an hour, if you are okay with that,' Kate told him kindly.

'Breakfast will be nice, ma'am, I just hope that it will not be too different in this time. I used to like bacon, beans and egg,' he replied almost salivating.

'Not entirely healthy, but I can get you that if you follow me to the dining area,' Kate nodded again so as to invite him to follow her.

They crossed a number of hall ways, well maintained with the fresh smell of paint, went down a stair case, very clean as well, until they arrived in front of this massive room with lots of tables. A few tables were taken by other people, dressed in similar gowns such as Vincent's, some people looked to be in good moods, chatting away, some others looked rather depressed, sitting on their own, deep in their thoughts perhaps. The whole scene wasn't very appealing to Vincent as he felt strange amongst all these people. He followed quietly in Dr Palmer's footsteps until she indicated for him to take a seat in front of a solitary table near the counter.

'Hello all, this is Vincent, he arrived here last night and he is going to stay with us for a while,' Dr Palmer addressed the gathering.

Some people looked up to her and then to him saying 'Hi Vincent', some others didn't look remotely interested, however the occurrence disturbed Vincent somewhat. He did not expect to be introduced and felt uncomfortable with it. He just took his seat and looked down at the table cloth, which appeared to be of some plastic material, freshly wiped, with green and white squares throughout. In the middle there was a small pot with a lavender plant in it. The smell of the plant hit Vincent's brain with vengeance making him have shocking visions of pure hell. It only lasted for a few seconds and it all went away as he shook his head and then pushed the pot with the lavender plant right to the other end of the table. Minutes later Dr Palmer went about her business while Vincent's mind wandered aimlessly, a plate of steaming bacon, egg and beans was placed in front of him along with a glass of orange juice and what appeared to be fresh coffee in a medium sized cup.

'Enjoy...,' this older lady, wearing a white apron, told him, as she disappeared back into the kitchen before he could fully see her.

Chapter Two

As he walked into the office Vincent noticed a middle aged man sitting behind an old heavy desk while Dr Palmer was standing aside, kind of leaning against the desk with her left hip. They were both quiet and their facial expressions were those of curiosity especially that of the middle aged man. Nonetheless they both encouraged Vincent to come in and sit on this green leather armchair that was pulled in front of the heavy desk. They did that by gesturing with their hands, both at the same time, strangely not saying a word. As Vincent sat down on the chair, the middle aged man got up revealing himself to be a tall man, not very broad but rather skinny, dressed in a brand new - or at least freshly cleaned and ironed to perfection - expensive suit. Vincent could recognise just by seeing the quality of the material however fashionable it was that it resembled nothing like the clothing gents were wearing back in his days. And this was in contrast to the office room which appeared furnished just like in his times, if you took away the strange looking machines that were lying around the place in no particular order. The middle aged man walked around the heavy desk and came right in front of Vincent looking down on him. He suddenly smiled displaying very white teeth. His face appeared freshly shaved while this discreet perfume started filling the air in Vincent's vicinity. The man's hair was nicely trimmed as well, his eyes deep green under a large forehead with thick eyebrows. In a few words his entire appearance was impeccable but not only this,

his demeanour expressed kindness and honesty. That was at least how Vincent perceived him at first glance.

'Hello and welcome Vincent, I am Dr Leary, Adam Leary, head of the psychiatry department here and Dr Palmer's boss so to say. How are you feeling sir,' the middle aged man asked in a kind voice, which confirmed Vincent's initial gut feeling about him.

'How amusing I must say, but in the last few days since I seem to have become cognisant somewhat, it is only doctors that I acquaint myself with, what is that saying about me?' Vincent responded in a caustic tone.

'Nothing else except that you need some help, some guidance so you can figure out what is happening to you. Dr Palmer and I are here to do just that. So you are in good hands Vincent.' Adam reassured him.

'Sure' am, but I don't see how you can help me, not after what I am about to say to you and Dr Palmer here, cause I have the sense that you already think that I am a dummy or a cuckoo of some sort and I can assure you that I am not,' Vincent said firmly.

'No we don't Vincent! At the most I can say that you are a little unconventional if you get my meaning. Um, essentially you are not the usual clients that we meet here day in and day out.'

'What is it that you need me to tell you Mr Leary, cause what you're about to hear is beyond your unconventional I am afraid, and that's not something I will be at ease to talk about, whether you are a doctor or not.'

At this point Dr Palmer moved towards the chair behind the desk and silently sat down, not losing eye contact with Vincent. She displayed a delicate smile and then picked a hardcover folder from the desk, which she opened.

'I think we shall recap your last few days,' she began speaking in a friendly manner, 'so we are clear of how you came to be in our care. Even so Vincent, pardon me having to say this, but it is bizarre,' she ended in a slightly irritated tone.

'I believe what Dr Palmer is saying is that we're a little confused here Vincent, cause you were not referred to us by our usual sources...,' Dr Leary completed.

Vincent just nodded, in order for the two doctors to fire their questions at him. In his mind it was as clear as day that these were just the wrong people to talk to about his story, but he had to admit that there was little to no choice for him. Perhaps if he told them whatever it was they were after, it would ease the way up to someone who really counted such as the president. But he also felt that reaching that high up, may just be a chimera, as it seemed that in this world in which he had inexplicably landed, things were being done differently. Back in his time an officer sent to accomplish an important mission would have direct access to the higher hierarchy and by no means would he have to be scrutinized by doctors. On the other hand he could see that these doctors appeared to be as confused as he was. Hence he thought that it would be best if he told them the truth, the only question remaining was, would they actually believe him? He strongly considered that what happened to him was unique, it was entirely unheard of and for these reasons Vincent felt very anxious about talking to anyone, but there again it was all up to him to be convincing, considering that whatever he was to tell them, was nothing but the truth.

'Okay,' he said daringly, 'you ask me whatever it is that you need to find out, I will answer, but have in mind that it won't be easy for you to understand for it is like a tale, but I will speak the truth as far as my own awareness will allow it, for I may have some memories that have faded away. It is not entirely clear to me what happened after I reached my endpoint, it has been like a fog covering my mind ever since but I do recall things, some horrible things and I also believe that I have been given a message by these "evil gods" so to name them, which I cannot summon up at this time, but I know it's important. They took me up there that I know...,' he ended pointing up to the ceiling.

'Um, I wouldn't go as far as to your "endpoint" Vincent or "up there". I would rather established a few facts at first, and then we will be pleased to listen to your account of the story if that's good with you,' Kate told him bluntly.

'Yeah, that's good with me, but bear in mind that I may not be entirely precise, especially if you raise questions about how I came to be found back in Brazil, that I hardly recall myself,' Vincent explained guessing the doctors intentions.

'That's fine, we may be able to help you with that at a later stage,' Dr Leary intervened.

'All right then, I need to ask you, is your name Vincent Monroe?' Kate asked letting her eyes down on the hardcover folder which she now placed on the desk.

'To the best of my knowledge, yes ma'am, that's my name.'

'Can I ask when were you born Vincent?'

'Eighteen seventy ma'am,' Vincent answered without hesitation. He then looked straight at Dr Palmer but saw no noticeable reaction in her facial expression, as if she expected that answer.

'You claimed that you are, or were I should say, a Colonel in the US Army, is that right?' she continued keeping her eyes on the folder.

'That's right and perhaps you were correct when saying that I was, rather than I am. I do grasp the fact that times have changed.'

'Can you tell us how you became part of the military Vincent? As far as you can recall of course…,' Dr Leary took over.

'It was a family tradition; my grandfather and my father before me were high ranking officers. I was in fact born and raised at Fort Brown in Brownsville Texas, since you did not ask, so I had adopted military fashion and discipline ever since I can remember,' Vincent told them visibly irritated.

'What was your father's name?' Dr Leary asked seemly ignoring Vincent's slight agitation.

'Alan, Alan Monroe, First Class Major, 20th Cavalry Regiment, during the Civil War that was and then later after being reinstated he served at Fort Brown. My mother Eléonore came from New Orleans and she was of French origin. My grandfather Dúghlas perished on the battle field during the Mexican war,' Vincent affirmed with conviction.

In the short interval that followed, Dr Palmer moved her attention from the folder she held close for a while to one of these strange looking apparatus placed next to a couple of more familiar looking objects on the desk. She then began performing some motion with her hands like type writing. A couple of moments later she lifted her eyes from this "thing", for which Vincent had no name still, and looked straight at Dr Leary expressing much surprise on her face, as Vincent could sense. She then suddenly turned the device around so that the luminous board faced Vincent as well as Dr Leary. In that moment Vincent felt how he caught his breath, his entire being becoming weak and nearly dropping off the armchair. He felt tears invading his eyes, instantly making the picture in front of him look blurry, his lower lip trembling out of control. Whatever words he had at the tip of his tongue just rolled back to the bottom of his throat, not willing to ever come out for it made no sense to speak these words, not after seeing the image of his own father shining up on the luminous board. The two doctors were stunned by his reaction too, neither of them being able to come up with anything sensible to say or do. The question "do you know this person" was indubitably inappropriate, total nonsense at this stage. For Vincent though, not only was the picture shocking but he was rather terrified that this machine would actually hold an authentic photograph of his own father. He felt lost, but comprehended the fact that the world he was in now had immeasurable knowledge. And then as if his mind was playing tricks he came up with a name for the amazing gadget, he called it "the knowledge machine". He looked over and back at the two doctors still incapable of putting his thoughts in order, so that he could say something reasonable about this development, but this paralysing sensation was still too strong to allow him do such a thing. He couldn't help but wonder if this whole new development was real, or else it was just part of the overpowering world of hell, where he had lived in pure anguish for so long. What could be more excruciating than being forever confused and uncertain about pretty much everything? Something though, perhaps the only rational bit still left, was telling him that he was very much alive and whatever happened now and happened before in a time that he could not define yet, was in fact part of his living experience and only God knew for how long it was going to last, and how much more there was to endure before something would end it.

'Vincent, are you okay,' Kate spoke, her voice shaky,' I take it that you recognised the man in the picture, please tell us about him.'

'That is my father, um, how you got his photograph I cannot grasp, but I can only guess that this wicked instrument of yours has loads of knowledge, like a knowledge machine,' he whispered.

'It's a computer, but you are right, it is like a knowledge machine, it's complicated. Anyway, have you not seen these "machines" before, like ever?' Dr Leary asked sounding very confused still.

'I've seen things, similar but not the same, and I have no name for them. Jesus there is so much that I need to tell, get it all out of me, things I've seen, painful things that happened to me, oh doctor you have no idea!' Vincent cried.

'We'll help you with that, I promise,' Dr Leary replied promptly this time,' but may I ask Vincent, have you been in battle during your time in the military,' he asked with genuine interest.

'Sure thing, I served in the Spanish War. I have been part of the Rough Raiders, training volunteers at first, then I was dispatched to Cuba were I fought in the battle of Las Guasimas and the San Juan Hills. I then went into the reserves and was reinstated to the rank of Colonel by President Taft, years after the war.' Vincent described briefly.

'I see,' Dr Leary whispered seemingly deep in his own thoughts,' that must've been tough. I propose to adjourn this meeting, it seems that you are quite emotional talking about this stuff Vincent, so a break will do no harm, wouldn't you say Dr Palmer?'

'Yes, totally,' Kate responded as if coming back to reality herself, 'Vincent we'll get someone to take you back to your room, I suggest that you take some rest and we'll talk more tomorrow, but just before you leave I need to check something out with you,' she said.

Without waiting for Vincent to answer, she got up and walked to a locker near the entrance and returned with an old backpack of unusually large size. She placed it on the desk and meticulously she began taking some items out such as clothing, an old pistol and some sort of leather covered notebook. Vincent sat quietly watching the woman although he felt another adrenaline rush. He involuntarily took a deep breath, betraying his emotions which were at surreal levels of excitement. In fact all his senses were acutely tuned to the present moment, so much so that he had little control over his entire being.

'Are these yours Vincent?' Kate asked after allowing a minute break so that Vincent could get over his excitement.

'Yes ma'am, they're mine,' he replied in a low voice.

'Do you give us permission to test them? I promise no harm will come to them; we'll only take a small sample of each item in order to establish how old they are. We call it carbon fourteen testing, it's very accurate. I also promise not to go through your notebook, not without you present, so what do you say?'

'That's fine by me ma'am,' he said not very convincingly as he thought the exact opposite, but then he realised how powerless he was over pretty much everything happening to him at this time. 'That knowledge machine of yours, very complicated as it is, holds like information, stories, photographs about pretty much everything there is to know, right?'

'Yes, that's correct. What you aiming at with that?' Dr Leary asked anticipating more shocking developments.

'Well, I can't explain how it does that but there may be a chance that a photograph of me is inside that thing. I took one right on the front steps of the White House before leaving for this expedition in the Amazon. There were a few of us from the group with President Taft, I am standing next to his right shoulder; can you look for it?' Vincent asked, his tone betraying much nervousness.

Once more, Kate's hands were producing the typewriting motion at much speed as Vincent could perceive, and then this clicking sound repeated itself at regular intervals.

'Um, there are a few pictures of the president that I can see, but not that many in fact, I have to search through, hold on a second I got something,....huh, is this the one?' Kate asked rotating the laptop to face Vincent.

The photograph was very grey, often crossed by dark shadows and while you could make out a group of five people standing on the steps, it wasn't easy to make out their faces except for the president's, which somehow was more in the light than the others. But Vincent had a clear recollection of that very moment, a memory projected by his mind in vibrant colours and also associated with powerful feelings, as they were assigned to a "historic expedition", as the president called it, in a place that a very few could reach back in that time. "The search for God" they were told, as these words struck Vincent's mind after staring at the luminous screen for a couple of minutes. He nodded unconvincingly, knowing deep in his heart that the quality of the photograph wasn't the best, although he could recognise himself in it. He lifted his shoulders, got up and headed for the door without saying a word to them. What was there to be said? If they were smart enough they would see it for themselves that he spoke the truth. If they weren't, well he would be labelled as a "cuckoo", making stories up. Unfortunately he had no control over what people thought of him and to make matters worse, he had no control over anything at all. It was as much a mystery for him how he came to be here, right at that moment; it was something beyond reason and it was something that he could not comprehend the mechanics of. So how could he expect these two strangers to even accommodate his account if he could not make any sense of it either? The only thing he could be sure of was that he was telling the truth. Moments later the two doctors were left on their own, very confused, hardly capable of contemplating what they were to do about this mysterious man and his mind-boggling story. They were confronted with delusional people every day, but Vincent appeared to be in a different league altogether. It was as if this man was telling the truth, something impossible to digest as indeed his whole tale was like a sci-fi thriller, so the second option was to explore his delusional state, which must be very deep. However there were a few things that the doctors couldn't make heads or tails of. For one they had to consider the location in which he was found and secondly the strange belongings that he claimed to be his. Furthermore they had to take into consideration his reaction to seeing his father's photograph that seemed way too genuine and then his suggestion to search for his own photo which was giving a new dimension to the entire case. Dr Kate Palmer was very good at her practice, working with people in high states of delusion; she was recognised for this, but for all the cases that she dealt with before, she had strong evidence to show them, so that the patients would eventually accept the reality of their circumstances. With Vincent she felt, it wouldn't be a smooth ride at all.

'What do you make of him,' she asked after several moments of pure silence.

Dr Leary was just standing in the middle of the room with his arms crossed over his chest in deep thought. He instinctively walked towards the armchair on which Vincent had sat, up to a few minutes ago, and let his body follow gravity, just falling on to it.

'I would like to see things from his point of view but that's impossible isn't it? I mean we've seen a few cases in our career, didn't we now? I just can't tell with this guy. Is he sane, just trying to make a big story up, but there again where is he coming from? He was found in the backside of the jungle, so how fucked up is that? A man coming from the Amazon claiming to have got there roughly one hundred years ago, sent by the US president of that time on an expedition, was taken "up there" by "evil gods" who gave him a "message". Was he referring to extraterrestrials, was he abducted? That's beyond hallucination, it's like a script for some fairy-tale beyond our wildest imagination. And then you have all these belongings that look to be from the past, his reaction in seeing a photograph of his old man or so he claims, and then he tells us to look for his own photos in our "knowledge machine". Is he acting all this up, I just can't tell!' Adam said with determination.

'I get the sense that part of you would like to believe him, like come on Adam. I have to say he is strange, perhaps the strangest we've seen in this clinic, but he is not that dissimilar to the rest of them. He is delusional, severely delusional, perhaps more so than others, but pretty much his case respects the same principals as many other people with this condition.' Kate advocated.

'Do you have a theory, something we could start from, Dr Palmer?'

'Yes I do, this man undoubtedly went to the Amazon, he was there for some sort of expedition, but it happened in more recent times certainly. Something dramatic must have occurred, he got lost, somehow he survived even for years, then he came across those items and he began hallucinating. To survive for a long time in one of the most unwelcoming places on Earth is drastic but he did it. Unfortunately the result of his adventure is his deeply troubled mind, it is the end result of PTSD, very severe I must say but it is what it is.' Kate pleaded nodding her head with conviction.

'I agree, the story is fascinating to say the least. Yeah it is the only possible lead that we can follow. We need to let him rest, for a few days at least, let him relax and maybe he will start remembering things more accurately. We will allow him to tell us his full version of the story anyhow and then only to try and help him see the reality. Meanwhile we get those items tested and I will search every possible American expedition to the Amazon for the past ten years, to see if any Vincent Monroe was part of it, if that's his name. Do you think the IT guys can run a few filters on that picture, make it more accurate?'

'I'll get on with it. I am somehow convinced that we'll get to the bottom of this in a matter of days, God I feel sorry for the poor fellow. I will report any findings that I come up with, to you directly and no one else. We need this to be kept strictly confidential, wouldn't you say?' Kate persuaded him.

'Absolutely but you need to let him know of this, in case he gets in contact with other people. I think there will be no harm in searching for any missing scientists, explorers and the like, for the past so many years. There must be something to prove his identity; we'll run finger prints, DNA testing, you name it. I shall not rest until I find the real Vincent Monroe, if there was one.' Dr Leary elaborated.

'Done deal, I will report to you the day after tomorrow unless some relevant development comes up…,' she told Adam.

'Yes, you do that. I will answer my cell phone even after hours if need be,' he replied with conviction once more.

For Kate though, Dr Leary's interest in the case seemed a little odd more so than any other time when they worked together on difficult patients. On any other case, Dr Leary would have been more pragmatic, more given to scepticism and with a very much more down to earth sort of approach. Now it appeared that his fascination with Vincent's farfetched story was getting the better of him. Somehow she felt that there was a lack of logic and professionalism in his perspective of this case. And she didn't necessarily base that on Dr Leary's conduct so far, but she felt that he was very tempted to research an implausible avenue and prove Vincent right. It was just a hunch she had, it was just the way she sensed his interest from the minute she presented Vincent's case to him. On the other hand she did not doubt Adam's ability to work on any case whatsoever, but she had to admit that Vincent's was a peculiar situation that would excite anybody's imagination, whether they were trained and experienced psychiatrists or not. Anyway this case had to be treated as unique, and it might turn out to have a simple, down to earth explanation, or it might end up in a long struggle especially if no immediate evidence of Vincent's identity was found.

Chapter Three

There is always something that constitutes the known world at any given space and time for anyone who is contemporary with that particular place in space and time. Take for instance the old roads, the new roads, the clean streets or the dirty streets, things that always change but somehow stay the same. And there are carts and horses, motorised vehicles, locomotives, vessels, railway stations and docks, all part of the same picture, the contrast between past and progress. Then one can add buildings, banks, saloons, houses, some new, others are old, some are clean and others are dirty, and this place, which is a town, is growing day after day and forever. But one cannot complete the portrait without including people. The good people, the bad people, young and old, males and females, natives and migrants, white, black, Asians and Mexicans, but they are all Americans. Some are tolerant, some others are not and there is a growing class that rises ever more powerful, in spite of being outnumbered. They spread fear and hatred, racial hatred that is, while others are being put back, becoming enslaved for they never were united, although they're far greater in numbers. Some are rich and others are poor, patrons and servants, some reaching for extravagance while others struggle to survive. It's what it is; the human jungle ever transforming, building blocks for the future, although nobody can have a mental projection of the future. It's what it is for now, the American society at its peak in the town of Brownsville Texas, in 1912. And in the middle of this entirely chaotic picture as it seems, there is Vincent Monroe, Reserve Colonel Vincent Monroe that is. He is now a drunk, working on the docks and taking refuge in the barn, on Clark's widow's farm, just outside the town, for she took him there out of mercy. But he cares about her and her younglings, helping out with whatever he could, around the household and then the farm during the season. Rumour has it that he used to serve Gina's other needs as well, womanly needs that is. She was still a young woman who inherited old Clarke's little fortune, after marrying him, only a few years before he passed away. Although having two children from her previous marriage, old Clarke took her in as a servant and then made her a praiseworthy wife. After his death she felt the need to return a favour to an acquaintance of her defunct husband, namely Vincent Monroe. Old Clarke felt sorry for Vincent, once a respectable officer and a gentleman, as they both served in the Spanish War. But then Vincent's life took a wrong turn and he fell into the abyss of his own sorrowfulness and shortly after he hit the bottle hard, so hard that at present he couldn't let a day pass by, without drinking his bitterness. He decided, it seems, to drink himself to death and there was nothing stopping him. There wasn't much for him after the war, he just returned to Brownsville and began feeling sorry for himself, didn't want to get settled into a family lifestyle, marry, have children and that kind of thing because of all the things he saw and did during the war. He often thought that he could not have children as he could not look them in the eye for he would only see the eyes of children he had killed, accidentally he was told, but they were children nonetheless. Why would he need a wife then? Not after he'd seen countless women being raped and murdered by his fellow soldiers. He was told after the war that there were no crimes being committed, they were just casualties of war, the wrong people in the wrong place at the wrong time, but that justification did not add up for Vincent. He thought he knew war because he was raised in that spirit by his own father, but it seemed that the wars his father and even his grandfather fought had a cause and a meaning. His war had none, other than the will of a few industrialists with expansionistic views. It was the only reason for which he and his fellow comrades were sent far away from their homeland, to fight, to kill and to commit atrocities and then fall in battle and become victims as well. It was upon his return that he understood the true meaning of war and once he did that he felt the weaker one amongst his father and grandfather, because no matter what was the cause for war, the horrors, the murders, the abuse were all the same in every war. Well, for that matter his own father and grandfather committed the same atrocities or at least they had been part of them, but then they had the power to carry on doing ordinary stuff, like having a family and so on and they even did it with pride and no consideration for remorse. Let's just say that Vincent was made of a different material, because now his soul was injured for life and he could not mend. He couldn't heal even with the care of his old comrade and friend, poor old Clarke, God rest his soul. Nevertheless, these days Vincent was all alone, working away at the docks for how long he kept himself sober on any given day and thankfully the widow, Gina Clarke, still let him sleep on the stack of hay in the barn, in exchange for the few jobs that he could help her with when he was sober. There was a time after old Clarke passed away when Vincent stayed dry for months and used to comfort her and looked properly after things, a time that she was even entertaining the idea of a companionship, but this was only short lived as Vincent went back to do what he did best, drink himself senseless every day. She hurt, perhaps even more than after burying her husband and now, God only knew for how long Gina was going to tolerate him coming near her house and the children. Like every other evening, almost, Vincent found himself wandering the streets, some new, some old, some cleaner than others, but always leading to the same place, the good old liquor place, "Crixell's Saloon". Deep in thought, as usual, he walked through the doors, stepped inside and stopped for a few moments just to contemplate the atmosphere inside the place. The saloon was busy at this time with all sorts of people, some poor, some rich, some better than others from whichever perspective you'd see them, but they were all here for the same purpose: to spend their earnings on cheap or else more refined liquors, beers and wines, and ultimately on the whores that were always keen to maximise their livings offering lustful dreams to many people, especially to those who could not find such pleasures anywhere else. Every time that Vincent walked into the saloon he found it amazing, how the rich were always hanging around with the rich, while the poor were just scattered around the place, hardly talking to one another, just eking out their pathetic existence with cheap spirits. He looked across to the bar and saw Teofilo, standing tall behind the counter scrutinizing the activity through the thick smoke. His place was usually peaceful due to his brother who was no other than Marshall Joe Crixell. For that reason, a very few would want to get into fights and then be arrested by the Marshall. He wasn't the most forgiving man, by contrast he was known as a very tough, hard-bitten individual, who would not hesitate to kill if need be. Rumour had it that his own deputy, P. McAllister, held grudges against the Marshall, but no one knew precisely why. However on that night neither the Marshall nor the Deputy were in sight, perhaps having to do their rounds, so Vincent walked through the crowd, not looking left or right, but aimed his sights straight at the bar, where his prize was awaiting; the cheapest whiskey, for he only had two tokens left before he got paid. He rushed to the bar, as he passed by a table occupied by five unfriendly rich local businessmen, who always provoked him, made fun of and shamed his name. He wanted to avoid this undesirable encounter and as they seemed busy playing cards, he thought he may just have a chance to get by unobserved. But that wasn't to be as two of them were facing the corridor between tables.

'Here comes "Captain Booze",' one of them said, while all the others lifted their eyes from their cards only to burst out laughing.

'He is a Colonel, mind you, isn't that right Vincent? What an ass, fucking utter dumb-ass,' another added.

'Leave him gentlemen, he ain't worth it,' somebody else, by the name Aaron told them without even turning to see Vincent.

'Hell no, he is worth it,' first speaker, Johnny Bruster said as he leaned back on his chair, 'he makes out with one fine lady, that Gina woman, poor old Clarke he considered you to be his friend and what did you do? You waited for the poor fucker to pass away, so you can pick up the goodies...,' he ended in disgust.

'By no mean he does that Johnny boy, I was told he tried it on with one of the hookers upstairs a couple of weeks ago, but couldn't get it hard enough, ain't that right Vince? You see too much booze and you can't aim your pistol not to mention you can't fire it.' Aaron laughed ironically. They all laughed making all sorts of foul jokes around the table.

'That's enough,' Teofilo thundered from behind the counter,' I suggest you go back to your cards, fine gentlemen and the ladies will be ready to join you shortly,'

'Now we're talking!' Johnny spoke with enthusiasm as they all returned to their game.

Vincent marched forward to the bar, only a few steps now, after his unwanted interruption. He didn't take much notice of what they said as he was used to their filthy comments, but it still hurt somewhat because he realised how his own pride was smashed to pieces. But he had to admit it was his own fault for becoming a drunk and a homeless, living at the mercy of Clarke's widow, when he used to be a fine officer and a gentleman.

'What can I get you Vince?' Teofilo asked as Vincent approached the bar.

'Same old, Mr Crixell, but look if you get me a bottle of Scotch for just one token today, you know I'll pay right? I always do,' he answered timidly.

'That's alright Vince, you're not my worst paying customer, I'll give you the home brew,' Teofilo told him while reaching for a brown bottle on the shelves.

Vincent nodded as if to thank him, took the bottle just handed to him and then picked an empty glass from the tray placed on the bar. He then placed the token on the counter and nodded again.

'Here, go to that table under the staircase, out of sight so to speak and keep it to yourself!' Teofilo said as he picked up the token.

Vincent nodded once more and headed to the dark corner under the stairs. He sat down on the solitary chair, poured whiskey right to the top of the glass and literally threw the liquid right down his throat in one swift motion. It hardly touched his tongue, however the warmth from the spirit invaded his mouth and then all his insides. It felt good he decided, and then repeated the whole motion three more times in the space of a few minutes. Now he was slowly coming to his "normal" state, as the beast in him was awakened once again. His mind, his vision and his soul were all attuned to one single being, crystal clear as he perceived it. Only for his darkest thoughts to surface again as brought back from deep within by the beast, and for that reason Vincent Monroe began searching his past memories, the killings, the pain inflicted upon others for it is killing that brands the beast in every man, when confronted with the brutality of war. He stayed there quietly for countless minutes, often looking through the steps above his head, to see the prostitutes' fine shaped legs and bottoms, as a yearning sensation overcame him, but simultaneously the clarity of his vision was fading away as he came to the end of his bottle. He was once again drunk, like the night before and the one before that, for as long as he could account for his insignificant existence. As he became more senseless with every last drop and his mind more tormented by the alcohol, Vincent did not even notice Johnny Bruster approaching right in front of him. At the last moment Vincent felt the man's presence, almost instinctively as if a small portion of his brain remained untouched by the strong brew, perhaps the small portion that kept some of his military training still intact, which for so long had kept him on his guard. And it was the same small part of his brain that told him to get up and stand still, while anticipating Johnny's next move so that he'd respond appropriately.

'Ok captain Booze it's time we settled this once and for all,' Johnny said with an intimidating air and without hesitation he launched his fist straight towards Vincent's chin however it missed its' target completely. It only took a small side-step from Vincent in order to avoid the punch, and even drunk his motion seemed effortless, simply because Johnny totally misjudged the distance between him and Vincent. As a result of his assault Johnny became unbalanced as his entire body went forward following gravity, nearly falling over the table. It was in that instant that Vincent used one sharp uppercut punch, which caught Johnny right in his stomach, making him bend and go down on his knees almost breathless.

'There is nothing to settle,' Vincent whistled his words through his teeth and stepped away from Johnny's body, which was now lying inert on the floor, just a constant groan to show that he was still alive.

Just a split second later Aaron and a few others rushed to the scene cursing, but were stopped in their tracks by the imposing appearance of sheriff Joe Crixell, who had been just inside the saloon for a couple of minutes now, enough time though to see what had happened.

'Hold it right there gentlemen, I take charge of this now,' he announced loudly as he positioned himself in between Vincent and the gang of lads approaching him.

'Well then arrest him! He assaulted Johnny for Christ sake ain't that right boys?' Aaron shouted back, obviously infuriated by the sheriff's inopportune emergence onto the scene.

'No can do gentlemen, I was only a few feet away and I saw Johnny trying to assault Mr Monroe here, only he missed him with his punch, but he attacked Mr Monroe in the first place. So you can pick your friend up, help him to his seat and get him some water, it may help him come back to his senses is what I say...,' Joe's imposing voice rumbled around the place.

'No can do? He is nearly dead for crying out loud! I don't get it Joe, who are you siding with huh? Are you with the "Blues" or the "Reds", because I got to tell you, it's hard for me to see you lasting much longer in your post Joe, rest assured on it.' Aaron told the sheriff while pointing his finger to his face.

'Stay calm gentlemen, I am the law enforcement in this town and I side with no one but the Law. Now to prevent any further injury I suggest you follow my earlier advice and I will deal with Mr Monroe, is that clear,' the sheriff added gravely, at the same time as he gently removed Aaron's hand away from his face. He then stepped back and pushed his coat away from his hip with his right hand, letting his pistol sheath come into view in a sign that he wouldn't hesitate to use the gun if it came to it. 'Now I need you to be wise about this Aaron Wright, help that waster of a friend of yours get up and go back to your table,' he then completed assertively.

'All right folks, just do as the sheriff says for now, our time will come I promise you that!' Aaron told his friends, trying to sound menacing.

Without delay they rushed over to Johnny, still lying on the floor, coughing severely now but also cursing something hard to make out. Once back on his feet he tried desperately to walk by himself but it proved rather hard, so he was almost dragged over to the table by his mates.

'That's what happens when you go up against a professional soldier Johnny boy, he may not be what he was but he still has it you know?' Deputy McAllister said in passing after watching the whole scene from a few feet away without interfering.

'Teofilo give Mr Monroe a bottle of that whiskey, put it on my tab, will ya? I think he wants to stay away from the premises over the next while, am I right Vincent?' Joe shouted across to his brother. 'Mr McAllister make sure that Mr Monroe sees his road home, we don't want any harm coming his way,' he said turning to his deputy who nodded instantly.

'I know my way Joe, thank you!' Vincent replied somehow becoming sober after the incident. He picked up his hat and took the bottle of whiskey handed to him by the sheriff and then slowly made his way out of the bar trying to look in control. A few curses were heard all around the saloon as people were getting back to their drinks, while the prostitutes were trying to restore the good spirits of the paying customers. Outside the fresh breeze stroked Vincent's face but had no positive effect, on the contrary it made his drunkenness come back with a vengeance. The road "home" was a long one too but didn't pose much of a problem to Vincent, as he was well used to travel in this state. He set his dirty hat back on his head and holding the bottle with much care he began walking, making desperate efforts to keep a straight line. A little while later he was just outside the town about one and a half miles from his destination, Gina's barn. He chanted all the way going from side to side on the dusty road, trying to picture just how ridiculous he must be looking but he couldn't give a damn, he felt happy. He became quiet as he got close to the farm, as per Gina's instruction to keep silent and not frighten the children. Within minutes he found himself inside the barn in pure darkness which only became deeper as he collapsed on the pile of hay. One by one all his senses switched off, as he went into his usual drunken deep sleep. After many hours of unconsciousness Vincent was slowly getting back to reality, sensing the cruel feeling of thirst combined with that of sickness. He wanted to throw up but abstained from such as he could hear many voices outside in the yard, all becoming louder as they got closer to the barn. He focused his vision on the spaces between the timbers, but could not see much as his sight was blurred from the massive hangover he was experiencing. Most voices were of unknown men but also Gina's could be heard. Not willingly this panic sensation took over his being and no matter how hard he tried to fight it off it set in, giving him a sense of apprehension. He got up on his feet anyway, trying to get ready for whatever was coming, but there was a problem. Vincent could not have even a faint idea of what was coming. His initial thought brought his mind to the fight he had in the saloon, just hours ago. Perhaps Johnny and his companions were seeking revenge and now they were coming after him right on Gina's turf, putting him in a hell of an embarrassing situation. It would simply mean that he would not set foot in her yard ever again whatever the outcome of the upcoming dispute. The heavy doors of the barn burst open with a loud screech, letting the sunlight invade the entrance. Vincent stepped back, instinctively covering his eyes with his right arm, but this action also made him stumble, coming down on his side. An imposing presence shadowed the entrance as Deputy McAllister stepped inside. Vincent attempted desperately to regain a standing position, as he couldn't yet see who the man was. He also distinguished two other men standing behind the first, which made him realise that he was in some trouble.

'Good morning sunshine,' the unmistakable voice of the Deputy echoed all around the barn. On hearing this Vincent's panic eased somewhat thinking that it would be better to be arrested than having to confront Johnny's gang.

'Did you have to sleep rough once more? Won't she let you inside the house, damn women,' McAllister commented sarcastically. 'Here, there are two gentlemen, a federal marshal and his deputy wanting a few words with you, honestly not a clue what they want, but I have the feeling that you are out of my jurisdiction once and for all,' he said coming down on one knee almost leaning over Vincent.

Vincent didn't react at all on hearing McAllister's words but he was rather curious to know what the federal officers had to say. It was rather important he felt, so he got up once again, keeping his balance a little bit better than a few minutes earlier in spite of the splitting headache he had to endure. He meticulously rubbed his palms against his clothing in an effort to remove the straws hanging off his dirty outfit. He then took a few small steps towards the large opening, covering his eyes with his right hand from the blinding lunchtime sun. The two gents in US Marshal's uniform nodded in a sign that he needed to step right outside the barn. Deep in his heart

Vincent was rather more confused than anything else, surely two federal marshals were not required to make an arrest over a trivial fight, he had the night before. Deputy McAllister stepped right behind him and whispered something in passing that sounded like a curse.

'Are you Vincent Monroe, Reserve Colonel Monroe, sir,' the younger of the two men in uniform asked politely.

Vincent did not answer straight away but looked over the two men to see Gina and her two children, who all appeared worried sick. This was exactly what he did not want to happen, to be arrested in front of the only people that he cared for. How embarrassing he was thinking, when he felt this firm push from the side as McAllister stood at his right shoulder.

'Can you not answer the deputy, Vincent? Of course he is Vincent Monroe,' McAllister spoke loudly.

'Very well sir, if you are Colonel Vincent Monroe please allow me to introduce myself. I am US Marshall Simon Martinez and this gentleman here is my adjutant Mr Hall. Sir I have a letter that was sent in by telegram here which I need to read out to you, so if you will listen carefully..,' this gent with dark skin and a large black moustache said, while taking a piece of folded paper out of the left chest pocket of his shirt. He unfolded the paper carefully and then grunted as if to clear his voice. In that instant Gina Clarke began running towards them shouting something incomprehensible.

'What did you do now Vincent, got into a fight I heard, an officer and gentleman you call yourself, shame on you bringing all these guards into my yard, you hear,' she shouted once more as she stopped close to the group.

'Ma'am we're not here because he got into some fight, not at all. So if I may I'd like to read this letter and once Mr Monroe is ready we'll clear off your lawn, rest assured,' Marshall Martinez told her, his voice annoyed. He then looked back down on his paper and began reading gravely: "I hereby by Presidential Decree wish to inform you, Reserve Colonel Vincent Monroe - last known duties in service to the United States First Volunteer Cavalry, until September 15 1898, last rank held Major, retired as Reserve Colonel- that you are reinstated as active. By orders from the Presidential Office, US Marshalls will escort you to Washington D.C. where a meeting is being set for you to meet with President Howard Taft, three days from today, April 17th 1912, in regard to your new assignment." This letter is signed by Secretary of State Philander C Knox on behalf of President William Howard Taft. Here it is sir, see it for yourself,' Martinez ended by handing the letter to Vincent.

It was hard for Vincent to even make sense to what he just heard. He peered over the letter and read it again word by word. And although it didn't look anything like a joke why in the name of God was the president reinstating him and further more wanting to meet him face to face? He shook his head gently and folded back the letter, then swallowed the lump in his throat and looked over to all those present. The astonishment on McAllister's face was immense, as Vincent could read, while Gina appeared rather passive, perhaps not yet assimilating the news.

'Sir you need to clean up, pick up a few belongings, and come with me. I only have three days left to take you to Washington and the next train to Dallas leaves in less than two hours,' Martinez announced briefly. He then turned around, saluted Gina Clarke and walked towards the gates.

'We'll wait on the road Mr Monroe,' the federal deputy said and then followed his boss.

Vincent stood still for a few more seconds, shook his head again as if only now he understood the message that he needed to be ready for a long journey ahead, all the way to the Capital. He went over to Gina first and gently placed his hands on her shoulders.

'I am very sorry if I caused any trouble ma'am, I just hope that I've been some use to you, but thank you very much for your hospitality. I hope to return soon if you'll still accept me,' he spoke in a kind, even regretful voice.

'Yeah, you have been of some use when sober but mostly not. Go with God Vincent and please do not return for both our sakes and the kids. Maybe you will try to stay off the booze and perform your duties with dignity, it sounds important,' Gina replied with not so much conviction. In a way, drunk or not, Vincent was fairly reliable for a widow such as Gina and he was a kind man, but with his mind clouded by alcohol most times she could not tolerate him for very long. She shook her shoulders attempting to escape his gentle grip and walked away to the children.

'You can use the bath tub in the house, I just boiled some water and take some clean clothes from old Clarke,' she shouted over her shoulder.

'One fine lady, um don't worry Vince boy, I will look after her while you on your mission from The President so go in peace,' McAllister said sarcastically as he walked by Vincent.

'You stay well away from this place deputy, you hear, well away,' Vincent roared behind McAllister, but there seemed to be no effect on the deputy who just kept on walking towards the gate.

Vincent never liked him, not one bit. The deputy was a cold hearted individual and unlike his boss, the City Marshall Joe Crixell, McAllister showed no empathy, no remorse and no consideration for others. Vincent always felt that McAllister would one day bring nothing but sorrow to the entire community of Brownsville. For now though it was just a feeling and nothing else.

Chapter Four

The journey to Washington wasn't an easy one, not for Vincent, mainly because he kept tormenting his mind with questions he could not answer, such as the meeting with The President and the fact that he had been reinstated in service and so on, but also his whole being was in a deplorable state. Being tired and not having the luxury to drink, not in the presence of two US Marshalls anyway, did not help matters at all. He felt greatly disturbed most of the time, sweating and getting severe headaches. Apart from all this his mind was going back and forth to Gina, whom he never said a proper goodbye to, but for whom he had much regret, for not trying to build a good relationship, a partnership as such. However it was all too late now, even if he was to go back to Brownsville shortly, she would never let him near her or the household ever again. At least so it felt after her last words to him and even though she didn't sound too convincing, the fact that she never showed up to say goodbye to him said it all. So Vincent had no choice but to hold on to faint memories of short pleasurable moments that happened in between his drunken rages. It was just one of those things that people never appreciate having, but then regret when it has all gone. And gone it seemed, right at this moment when the steaming train came alongside the platform, slowing down with each puff until it came to a stop in the Capital's Union Station. Slowly people from Vincent's car made their way to the exit doors while he and the two Marshalls waited patiently until the corridor cleared. Within minutes though the three men where stepping out on the platform in the fresh air of the early morning. It had been a while since Vincent was in the Capital; in fact it was almost twelve years from the day. But the changes in the architecture if any were of no interest to him because now the sense of anticipation was taking over. In a couple of hours just, he would be face to face with the head of the United States and that was something you didn't get to do every day. Overwhelmed and nervous because of all this Vincent got into the Model T Ford, provided by the Washington Police Department, and didn't even notice the length of their journey to the White House. There were just a few words exchanged between him and the officers, which helped Vincent focus on the meeting ahead. But with no indication as to what this meeting entailed Vincent could only speculate on the importance of it. The fact was that the President would only call upon retired personnel once in a blue moon, but to meet retired officers face to face was extreme. However he was here now, right inside the perimeter of the very settlement that hosts the President, as the Model T roared its engine into the driveway, after a short pause at the gate. The two Marshalls only escorted Vincent as far as the first office, after passing through the sumptuous entrance hall, filled with paintings and expensive mirrors as well as candelabras hanging from the high ceiling. Here a middle aged lady greeted him and the federal officers and then showed them out after thanking them for their service. They were also asked to sign some forms just before departing. As she came back in she invited Vincent inside some sort of anteroom and asked him to sit down in a comfortable armchair placed in front of a nicely hand crafted coffee table.

'How was the journey Colonel, if I may ask,' she spoke as soon as she saw Vincent seated.

'Long ma'am, it was painfully long and when travelling with people who lack talking skills it seems even longer, if you get my meaning,' he replied emphatically.

'I sure do,' she said with a faint smile,' would you like some breakfast,' she invited him. Vincent nodded with some conviction given the fact that he was starving at this stage.

'I will organise some bacon and eggs right away sir, on the chest beside the divan there, you'll find some of the morning papers' she told him pointing to a two seater sofa located by the far wall. Another piece, nicely crafted in the form of a hardwood chest was placed along the side wall, holding a tall lamp and several newspapers. 'I will return shortly. Coffee or tea with your breakfast sir?'

'Coffee will do just fine ma'am, sorry I didn't get your name...,' Vincent replied pleasantly.

'Gabrielle, Gabrielle Lindergart, I am a protocol officer around here and for your information your meeting is set to take place in an hour's time, I will be back with your breakfast meanwhile,' she responded promptly. She then made her way out of the room and closed the big wooden doors tightly behind her. She wasn't to return for a good twenty minutes and when she did she was accompanied by a Hispanic looking young women, dressed in immaculate white apron upon a blue dress. She was holding a silver tray covered by a silver bowl, which she placed on the coffee table in front of Vincent. She nodded without a sound and then disappeared through the large doors in a flash.

'Please enjoy your breakfast Colonel and if you need anything else, let me know. I am just next door,' Gabrielle told him before making her exit and once more pulling the two doors tightly behind her.

From that moment time flew by and without taking much notice Vincent found himself walking behind Gabrielle through large hallways, beautifully decorated with English wallpaper in all sorts of floral patterns. Shortly he arrived in front of what might been described as a meeting room and was invited to follow Gabrielle in, after she had stepped inside and then to the side allowing him space to walk by. Four men were seated, three of them on two sofas opposite to each other, while the other, an older big man was sitting behind a heavy desk. The aroma of fresh coffee and Cuban cigars struck Vincent's nostrils the instant he walked in.

'Colonel Monroe. A pleasure to finally make acquaintance with you. I must say that Teddy had nothing but laudable words to say about you; he highly recommended you as a reliable gentleman, courageous but level headed and the like. I believe you two served in the war some time ago and therefore this information was given to me first hand, so I cannot doubt Mr Roosevelt's trust in you,' the older man from behind the heavy desk said while stepping towards Vincent in slow motion. He was a weighty man and appeared to be in his fifties, but one could not tell precisely, with dark hair nicely combed to the back and a white, nicely trimmed moustache. He was wearing a dark grey English fabric suit, expensive looking and perhaps custom tailored for him.

'William Howard Taft, President of the United States and these are Professor Alexander Englewood from the Royal Society, Philosophy Professor Alan Matthews Junior from Harvard College and the gentleman sitting on his own is Dr Frederick Jacobson, a private practitioner. Please make yourself comfortable Colonel, for we have to talk about some very intriguing matters,' the president said gently tapping Vincent on the shoulder indicating him to sit on the empty spot on the divan right beside the doctor. Without hesitation Vincent took his seat almost suffocated by a mix of curiosity and anxiety after realising that he was meant to be involved in this gathering amongst two scholars and a doctor. What would an ex-military man be doing here, in front of three men of letters, more or less, and the US President who called them to discuss "intriguing matters."

Perhaps this wasn't about war and, for all Vincent knew, in spite of his alcohol influenced blurry past few years or so, America was very much at peace. He desperately tried to hide his emotions and made himself look detached as all the other men showed no signs of nervousness.

'Gentlemen, just to set the ground rules for our discussion, I need to say that we'll carry it out with strict discretion, let's just say, and I hope you will all agree, that what we are about to say here may bring the Presidential institution into an awkward predicament, I am sure Colonel Monroe will concur after hearing this, is that clear?' the President asked inquisitively.

As they all kept quiet, signalling that they understood, the big man went back to his desk, and in slow motion once more he allowed his body to slide into the luxurious green leather armchair. He then raised his hand towards Dr Jacobson and nodded gently in sign that he should begin the conversation. The doctor grunted at first and lifted himself off the sofa.

'Yes, thank you Mr President, and with your acceptance I would like to brief this small gathering on some events that I came across not long ago. And if the two professors and indeed you Sir are already familiar with some of this, Colonel Monroe is hearing this for the first time, so gentlemen pardon me if I repeat myself. About three months ago I met an old patient of mine, an explorer of Scandinavian origin, who was miraculously returned from an expedition he carried out in South America,' Jacobson began his story in a rather casual tone. It was obvious that he had repeated himself on a number of occasions now. 'Mr Sigmund Rasmussen, that was his name, may he rest in peace, for he passed away last month, came back to the US after going through some formidable events on his venture to the Amazon. He appeared to have suffered from some unknown illness and, trust me gentlemen, I have seen some bizarre diseases in my twenty years of practicing medicine, but this was just different from the all others combined,' he explained, his voice becoming deeper and actually more enigmatic by the second, 'um, to my embarrassment I can't explain this any better, but the man seemed to be consumed by some sort of cancer which was burning him from the inside out. He also developed leukaemia as my tests show and a serious malfunction of the thyroid gland for which I found no reasonable explanation. His skin became greatly infected for which again, I am not able to offer an explanation. It has to be said though that the entire illness lasted for several months, considering that he travelled back to the States for almost eight months before he died.' Dr Jacobson elaborated but sounded more confusing with each word. More so for Vincent who was listening to all this without being able to make any connection between his requested presence and the sad story of some Scandinavian explorer, who happened to have died due to an unexplained illness. He remained quiet hoping that enlightenment was on its way. And as the short, bald Mr Jacobson took some sort of a notebook out of his briefcase, enlightenment was not far away indeed.

'Well gentlemen if Mr Rasmussen's death sounds odd, what he noted in his journal, which he gave me in good faith that I would report it to the appropriate people, is even more peculiar. He describes, "it is possible that I found God in this remote part of the world, for I have no other explanation for all the things I saw, after I was stranded on a plateau, well above the natural tree line of the jungle that I came out of, only a couple of days ago...." That's what this paragraph says,' Jacobson remarked after closing the hardcover notebook which he then set on the president's desk. He then returned to the middle of the office, took a deep breath and shook his head as if to regain his thoughts.

'There are more descriptions of things that are hard to imagine, flying orbs, pyramid like structures, energy balls, floating large objects at the edge of the sky and so on, no one can tell what Mr Rasmussen had witnessed but it was something that I have no doubt triggered his extraordinary sickness and ultimately his death,' Jacobson concluded with a slight moan.

Vincent couldn't make heads or tails of what he heard, moreover he wasn't even sure he understood the language in which the doctor briefly explained this strange occurrence involving the explorer of Scandinavian origin, however he decided to wait for more information before asking any questions. Still he could not determine his role in this conversation which was now becoming heated as the president invited the two scholars to share their view on the matter.

'It appears to be out of the ordinary indeed, but surely there must be some logical explanation for this incident that unfortunately brought Mr Rasmussen to his death. However I can't see how one can invoke God in all this, it just doesn't make any sense,' Professor Matthews explained sounding mystified, although it appeared that he had heard the story before.

'I am not so convinced that my dear colleague is in fact entirely right about God not making any sense in this. Now I hardly intend to say something that sounds insane, but I carefully looked through Mr Rasmussen's notes and some of the accounts given match a theory that I have proposed over in England. I and few of my associates at the Royal Society are contemplating the idea that God should, for a change, be taken out from the theological viewpoint and perhaps be given a more scientific outlook. All major ancient cultures that we looked at, the Egyptians, the Greeks, the Romans all the way to the Indians, are talking about their deities as a tangible presence, beings descending from heavens, Mount Olympus, you name it, but they are nothing like the mystical God that Christianity made God out to be,' Professor Englewood expounded with determination in his classically educated King's English.

'Outrageous idea, excuse me, but you cannot evidence God's existence based on ancient mythology...,' Professor Matthews nonchalantly chastised the Englishman.

'Alright, there is no need to enter this dispute now gentlemen, how about you Colonel? What do you make of this? What's your belief when it comes to God?' President Taft asked politely.

'Well..., I haven't given it much thought honestly, not in a long time, but back in the war, I did not see much indication of Him. On the contrary I felt that He abandoned Man and left it all in Man's desire and authority, which made me wonder back in those times, has He ever been a presence in our lives at all? Why teach Man all these things, good things, and then just leave him to do the exact opposite? I don't know, but something else strikes me. Did we make Him up so we can justify our crimes and then ask for forgiveness?' Vincent explained, sounding confused and even nervous.

'Are you a nonbeliever, Mr Monroe?' Professor Matthews inquired.

'I am just lost Professor, that's all. Things I've seen and things I've done do not fulfil what I was taught of God.'

'Nothing but the truth Colonel, as a matter of fact we are all lost and perhaps explaining God is not a simple task, it never was. I am a believer, but having said that I must agree there are nothing but questions,' President Taft concluded.

A short moment of silence followed, which gave Vincent the time to loosen his thoughts after this ambiguous conversation. And at the same time he just had to raise his own question once more; for what reason was he a part of this meeting? Why was he called amongst scientists to answer questions about something so profound that not even they could answer? It was nothing but a cause for frustration, not to say that it all appeared surreal, the president, the scientists, and him talking about God, based on a claim that some explorer had made, to have found God. It just didn't make much sense.

'If I may tell you gentlemen,' Vincent articulated unwillingly, as if his internal voice came out loud, "Um, I feel more lost not knowing why a man like me was invited to partake in such a deep discussion, I am just a retired officer, well until a couple of days back I was, but still what meaning has all this got for me, I hope you don't mind me asking,' he ended.

'We'll come to that in just a moment Colonel, but before that you may want to take a look at some depictions that Mr Rasmussen crafted in his notebook. You don't need to try and explain them, we couldn't, but more for you to grasp what I am going to further expound. Mr Jacobson do you mind handing the notebook over to Colonel Monroe?' the president asked politely.

'Not at all,' the doctor replied while picking the notebook from the desk. He then walked across to Vincent and passed him the nicely leather covered and rather heavy book. 'The drawings are towards the end. What you're about to see looks even more bizarre than what you have heard, I can assure you,' Dr Jacobson said, his voice enigmatic.

Vincent opened the book and, as he was told, flipped the pages carefully until he reached the end part of the document. There he found maps, drawn in charcoal with names of what appeared to be rivers, locations and so on. Farther down the pages there were depictions of plants, birds and those of humans in strange clothing. These pictures were drawn in colour and appeared rather well detailed. Then Vincent went through a few pages with symbols, which were hard to establish what they meant, however a few notes in English were made on the side of each page. Finally towards the last few pages pyramid like structures were depicted on what seemed like a large plateau, downhill from where the Scandinavian explorer was observing them. At the top of the structures some "fire balls" were portrayed in powerful yellow crayon while above these some grey looking oval objects were seen, appearing to hang off the edge of the sky. The word "gods" was used at the top of each of these pages and some other inscriptions, some hard to decipher, but also the words "temples of the gods" could be read. Vincent found this rather confusing and not at all something that he could make sense of, however he decided to wait for further enlightenment on the issue before he would try and build a rational picture of what he was seeing. He closed the notebook and nodded as if he was ready to hear why he was part of this conversation after all.

'All right then,' President Taft began 'I've got to tell you that I firstly didn't give it much thought either, in fact only because I knew Dr Jacobson to be a trustworthy gentleman did I consider this story for further examination. And in doing so I addressed this matter with a friend of mine, well in spite of what people think, we are still friends, no other than Mr Teddy Roosevelt..., um he is a man with a large range of interests as you well know. He only said this to me "Well Howard if I was still in office I wouldn't hesitate to assemble a search party to go out there and explore these findings", and here I am gentlemen. I am willing to do just that, so I need your help and especially your help Colonel, you see Teddy spoke highly of you and the times you served under his command in the Spanish war,' the president said leaning back in his sumptuous chair.

Vincent's senses were really excited now, he could hardly establish whether it was for the better or for the worst, but he surely did not expect this. Was he to be sent to the Amazon in a search for GOD? It sounded absolutely insane but it seemed that's what the president had in mind. But still, why him? How had Teddy Roosevelt described him, so that he was called in to what appeared to be a set up for an expedition deep in the rain forest of South America. As far as he could recall he was hardly distinguished by his former commander for he did not do anything heroic, nothing that would be classified as outstanding. On the contrary Vincent fought his own battle during the Spanish War, a battle to overcome his owns fears as he struggled to stay alive at all costs. He never thought that he did anything remarkable except whatever any other man did back on the battlefield. However he was here now to hear a statement that he found not only hard to believe, but it was coming from no other than the US President. It would be pointless to try to search for reasons, he thought, and it would perhaps be wiser if he waited until the President finished all that he had to say. The fact was, he had not yet called upon Vincent to be part of this search party, but he expected that he would.

'Um, gentlemen if you follow me, I have something extraordinary to show you in a more confidential place,' President Taft spoke again as he lifted his heavy body from the luxurious armchair, 'but before that I need your acceptance of this assignment. This is on voluntarily basis, no one is obliged to go, but it would be great if you all would. Additionally, Colonel Monroe can perhaps assemble a protection squad, made up of a few of the veterans that he fought with side by side in Cuba, experienced men when it comes to that wild sort of environment, maybe up to five or six people,' he concluded, and then stood still as if expecting all to reply instantly.

'Have we given enough consideration to this, gentlemen? Because I tell you going to the Amazon is nothing like you have ever heard of before. However we have to admit that Mr Rasmussen, although defunct now, made it back on his own,' Dr Jacobson explained.

'It is a bloody hell of a claim, personally I am in,' Professor Englewood said strongly.

'Yeah, I am too, I think it is an unique opportunity to maybe find something unprecedented indeed, although I have my reservations when it comes to God, nevertheless it sounds remarkable if it is true.' Matthews confirmed as well.

'I never thought I'd be ask to be part of such a thing but you all need a doctor, so I am coming,' Jacobson told them.

'Mr Monroe, Colonel, will we count on you to accompany and even lead this expedition?' President Taft asked patiently.

'Sir, fine gentlemen, I don't mean to be awkward,' Vincent begun nervously, 'but based on what I have heard, and I am mindful of not being an expert, how certain are you that Mr Rasmussen, God rest his soul, saw all these things? Could it be that he saw something of the ordinary sort but he interpreted these things as out of this world?'

'I think not Colonel. You see he was an experienced explorer and a scientist, he would not be baffled by anything commonplace, and neither would he put his good reputation to shame by making it all up. Besides he took a massive risk that eventually claimed his life, something that he wouldn't do lightly in spite of living in the wilderness most of his life, he was a very cautious man. I knew him well..., however Colonel we need to be well prepared and not rush into this, if you'd join us we trust that with your experience and with a squad of men like yourself, we've got a good chance to run this expedition smoothly and return all well from it.' Dr Jacobson suggested.

'Mr President, Sir, I don't know any men that would volunteer for this, even if I am able to find them,' Vincent cried.

'The remuneration is substantial I can assure you,' president Taft replied instantly.

'In that case I think I can put up a list, but I have no idea where to look for them,' Vincent said. However his thoughts were struck by the one word, remuneration, as he never thought of money being a part of this mission. However he now had the President's assurance and he could not help but think how better off he would be with a bit of money. And then how proud Gina would be, perhaps, if she saw him after doing something commendable for a change, and moreover he would be able to repay her for all her kindness. His thoughts were cut short as the president grunted loudly.

'Excuse me gentlemen, but this is an exciting moment now that Colonel Monroe has agreed to join this historical expedition, I take what you said as a 'yes' Colonel am I right?'

Vincent nodded but felt powerless to say anything. He was experiencing something in between agony and ecstasy.

'...., well then, as soon as we get the preparation under way, and as we all agree, it has to be well thought out, then you'll find yourselves on this rather exciting quest.' President Taft concluded.

A brief moment of silence followed, and Vincent used it just to search his mind for names of his old comrades in arms. Seconds later he already had a few in mind; the question being, were these folks still about? He instinctively walked to the president's desk, picked up a ballpoint and a piece of blank paper he found lying there and started to inscribe the names on it.

'Maybe the US marshals can pin these people down, bring them over and I'll speak to them,' he said after handing the note over to the president.

'Will do, Colonel, I like fast thinking. Now gentlemen if you walk with me for I promised to show you something,' the president announced as he made his way out of the office.

After a short stroll down the luxurious hallway, the four men entered a large room, like a library, as they followed the President. The atmosphere inside was sombre, filled with the heavy scent of old books while the illumination was very murky. President Taft walked over to a larger desk than the one inside his office, where he turned on a lamp sited on it. On the desk laid this painting of what appeared to be no other than the portrait of the founding father, George Washington. But this picture was odd, nothing like any other portraits that the public would have been used to seeing.

'Gentlemen it is this painting that I wanted to show you, and before you question, yes I think it has got something to do with what we talked about. If you would like to take a closer look you will notice this luminous, orb like object, just above good old George, there on the left hand side, upper corner,' the President said, pointing out with his finger an area of the picture where a yellow illuminated disc could be observed. However the painting was strange in every other aspect, where other symbols and pictograms could be seen as well as the inscription "Washington Freemason" could be read on the arcade above the founding father's head.

'Was he part...,' Professor Matthews attempted to ask, only for the President's severe voice to have him swallow his words.

'I won't comment on such things, fine gentlemen, I would rather draw your attention to the object I showed you and perhaps you'll find it difficult not to notice the resemblance between it and Mr Rasmussen's depictions, wouldn't you say?' President Taft commented, his voice echoing through the entire library, 'you see, when I looked at this painting after seeing the pictograms in the notebook, I was quite shaken by the similarity. I can't imagine how or why this is, but I can assure you that our Founding Fathers believed in something out of this world, and that, gentlemen, convinced me to grant this expedition,' he finished.

The four men kept staring at the painting and had to agree that there was a striking similarity with Rasmussen's drawings indeed.

'Gentlemen,' President Taft uttered,' I will call this meeting adjourned, but before I do that, I have one last announcement to make. Once you return from your quest you are to report only to the President, whoever that may be. I will inform him about this, and I will be informed in return about your findings. I once said that we will try to accomplish as much as we can without the drama, referring to the popularity Teddy enjoyed while in office, but I will tell you now, there will be no noise about this whatsoever, is that clear gentlemen?'

They all nodded silently, their minds wondering about this future expedition that they were all to be part of. What sort of adventure it would be, dangerous without a doubt, what sort of findings would they come across and what answers were they to give, to perhaps the next president of the United States?

Chapter Five

There were the blue skies, the blue sea and an everlasting undulating motion accompanied by occasional sickness. And there was also a cloudy mind and a million questions filling it, for there were no answers to be found, not yet, not in the middle of this borderless ocean, but the time was short before stepping on dry land again, but what for? If Vincent found the journey by sea to Brazil was difficult, how was he supposed to carry out the entire mission and go all the way to the far side of the vast rainforest, to find "God"? And what guaranties were there that he would reach the endpoint, right through the heart of the vastest jungle in the entire world? Must he have been insane to even entertain the idea of taking on this assignment? That may be so, but when the assignment was coming from the US President directly, it was perhaps hard if not impossible to turn it down. And if this argument was enough to justify his presence aboard the "Elvira" steamboat from Florida to the Amazon, he had to admit that he rushed into it and did not consider his own life to be at risk here. He may not ever see the borderless blue ocean with its' infinite blue sky, not that he'd miss any of it, but no one could tell for sure that he and his crew would ever return from this expedition. So what then? Not a lot, but just a glimmer of hope that they would be back in the States in one year, even eight months if God had mercy on them, or at least that was the plan. They were now near to the mouth of the Amazon River, the largest estuary in world, the place where this amazing body of water brings down and dumps all the secrets of the jungle to forever be lost in the dark depths of the ocean. Once the steamboat "Elvira" entered the river's mouth there would be a few more days, sailing upriver until the town of Manaus was reached, on the right shore of the Rio Negro, an eleven mile detour from the Amazon itself. From then onwards nothing much would change as they would sail back onto the Amazon and farther on upriver, the only difference would be that they would sail on a smaller boat. Slowly but surely the vast ocean was being left behind as "Elvira" entered the estuary and then after another while the shore on both sides began to close around the steamboat. "Elvira" was a small to mid-sized vessel, capable of sailing on both large rivers and the sea. It had a small crew of only five including the skipper. But aboard there were seven more men along with Colonel Vincent Monroe who had been given the task to lead this expedition. There were two professors and a doctor while the rest were ex-military, Vincent's old pals whom President Taft needed to find and recall to service. For these men, only the twenty thousand dollars each cash reward swayed them to sign up, or else they would have never pondered it. Even so they were hard to persuade, as Vincent carried out talks for days, but when they saw the five thousand dollars cash advance in their payment, there were no more complaints for cash always speaks louder. Vincent too was pleased with the money which he quickly sent over to Gina, figuring that five grand would potentially get old Clarke's farm back in shape. It was the only thing to please him really, as everything else about this expedition he thought of as being insane. And for that matter he could not relate to the professors whatsoever, as they seemed to be overly excited, talking out loud about science and theology. Vincent had received formal education, good education in fact, but even so he found it hard to decipher what these men were talking about. Every so often some of his comrades would ask him sarcastically if the professors were speaking English at all. He wouldn't necessarily excuse their humour but for the sake of keeping them in good spirits he'd answer likewise, using comical remarks. The fact was that Vincent no longer knew these men. After the war they went their separate ways and they hardly kept any contact thereafter. But now it was almost déjà vu in lots of ways, as Vincent needed to get to know these people again, get close to them and assume his role as a leader, same as before, at the start of the Spanish War. It may have seemed different, it wasn't war they were heading into, but he somehow knew that death was there and surely not all of them could avoid it. For that reason he found it imperative to be honest from the start and spell out exactly what they were getting themselves into, in spite of the big reward. They seemed to have acknowledged the risks and sadly some of them were even thrilled to know that their families were well looked after if they were not to return. In any case at least these were trained professional soldiers, well capable of fighting for their lives, but the professors and the doctor were pure novices when it came to survival in harsh environments and even worse, they appeared to ignore Vincent's talks and training during preparation time. Now they were only a short time from their first point of arrival, the port town of Manaus. It still felt agonising, for Vincent especially, to be on this boat, but at least there was no more sea sickness, sailing on these waters felt smoother. He had to admit that not drinking any alcohol for over a month now wasn't helping either. With the lack of any other activity aboard Vincent retired to his cabin and decided to begin his journal, something he intended to do ever since they were out at sea, but couldn't due to not feeling well. He went right in after dinner, after being told by the skipper that they should dock early the next morning. He asked his companions not to disturb him for the night and advised them to get rest, as from tomorrow this could become a luxury. Once inside his compartment he made himself comfortable on the wooden chair and opened the leather covered brand new notebook, which he received from the President. He then placed it on a little square table that he had in the small cabin and began writing.

Colonel Vincent Monroe's Diary

First Entry; May 28th, 1912. We were at sea for over ten days after leaving Florida, but now thankfully we reached the Amazon River and my wellbeing is much restored since the ship sails on these calmer waters. I was told that we'll reach our first point of destination early tomorrow and I have to say that I am looking forward to get off this boat. My companions seem to have experienced the better of this journey so far and haven't been affected by sickness. I and seven people are apart from the "Elvira" crew, which only counts for five men. We sailed into this part of the world as a search party to take account of some strange phenomena that were recorded by a Scandinavian explorer at the far end of the jungle. Therefore my crew is made up by two college professors, an American, Mr Alan Matthews, an English man Mr Alexander Englewood and Dr Frederick Jacobson from D.C. I further recruited four men for this expedition, all former Army personnel, Reserve Lieutenant Wade Johnson, Reserve Sergeants Clive Orson and Ramon Gonzales and finally Reserve Corporal Ahote McNealy also known as "Running Wild". They're all in good health, which is much needed before the start of this adventure. I much fear and it is my strong believe this is a very hazardous mission and I cannot guarantee the return of all these people, including myself. I hereby commit to write in my diary every time I find a chance, in order to try and record all the events as accurately as possible. Signed: Colonel Vincent Monroe

The fog was slowly lifting the next morning, allowing visibility to grow clearer with each yard the steamboat was gliding towards the dock. The town of Manaus was also taking shape in the distance from down on the water level up to the gentle hilltop, showing grey rustic buildings and every so often, some more colourful ones too, but they surely were all very distinct from the deep green of the rainforest. It was a bizarre site in many ways, almost ferric in spite of looking quite poor at first glance, but to see such a large settlement this deep in the jungle was certainly far from common. Although it was established a couple of centuries ago as a commercial port, the town of Manaus only grew bigger in more recent times and like elsewhere this town was the end result of progress, brought here at the start of the rubber boom. Not only did it grow in size since, but there was nothing to suggest that this port was isolated and not part of the modern world, not when it had electricity, villas, yachts and even an opera house. Along the shore many commercial establishments could be observed, especially around the docks. And it was one of these establishments that the eight men aboard the "Elvira" were seeking to find, "Thorn & Son Ship Hire and Repairs Ltd". The company was owned by an American, after buying stakes in a British owned firm that nearly collapsed about a decade back. Rumour had it that Frederick Thorn, the sole owner of the company lost his only son to the jungle, shortly after they settled here, however he kept the initial name of the firm in memory of his son. As the boat slowly came to dock, in the frail morning sun, a large pale blue work shop came into sight, with a visible inscription "Thorn & Son" on top. Vincent advised his companions to follow him but also warned them that he would do the talking. First they took down their gear from the "Elvira", which was quite significant in volume, but as they organised it well on the esplanade, it didn't look like a lot. They left Running Wild to guard the goods as they walked towards the warehouse.

'It's a bit early; do you think he is up working this time of the morning?' Wade Johnson enquired.

'Nope, but we can only find out.' Vincent answered, his voice neutral.

'It's great to be back on dry land anyway, so we can wait.' Clive Orson added from behind.

Indeed, Vincent thought, it was a blessing to be able to walk for more than 30 yards on something that felt stable for a change. As they approached the building they observed that one window at the front was lit, in what appeared to be a small room, an office perhaps.

'Well, the window cleaner has been relieved of her duties it seems!' Ramon Gonzales commented pointing at the thick layer of dust covering the window.

Vincent kept on walking, ignoring any of the comments. He had more important things to focus on and was rather looking forward to seeing this business through, to acquire a small boat, find a place to crash comfortably for a day and then have his crew on their way farther on, up the Amazon River. It should be straight forward as he was told that once he reached Manaus the boat should be already hired, equipped and ready for the journey ahead. The President assured him that he'd look into it personally, so he must have already wired the money along with a telegram to "Thorn & Son" announcing their arrival. At least that was the plan and in theory it should be simple, but somehow Vincent felt apprehensive about it. Without notice he found himself climbing the few steps outside a wooden door which held the inscription "Office and Registration". He gently knocked and waited for a quick reply as he thought that as the light was on, someone must be "home". Indeed just a few seconds later the door opened before him, revealing a man of medium height, probably in his late fifties, with the appearance of an individual not very well looked after, not necessarily dirty but rather scruffy. He was bald at the top of his head with some greasy grey hair on the sides while holding a pipe on the corner of his mouth. He meticulously took the pipe out of his mouth displaying a rather curious expression on his entire face.

'What can I do for you gentlemen at this early hour,' he enquired in a deep voice, desperately trying to look over Vincent's shoulders to see the other men.

'My name is Vincent Monroe and I think you should be expecting us, if you're Mr Frederick Thorn.' Vincent replied promptly.

'I am indeed, at least as far as I can remember, but what makes you think I was expecting anybody,' he asked with an air of sarcasm. 'Come on in, Mr Monroe, find a place to sit, perhaps you are here for some business, since you knew my name.' Mr Thorn invited him as he walked ahead.

Vincent briefly turned to his companions and signalled them to wait outside and then he followed Mr Thorn inside the room. It appeared to be an office alright, but felt stuffy and it was certainly overcrowded by all sorts of folders, files and papers, rolled maps and the like, thrown chaotically all over the place. However there was a desk with two wooden chairs on each side. Respectfully Vincent allowed the man to take his seat first and then he sat opposite to him across the desk. A heavy register was placed on the desk beside a gas lamp and some writing tools. A single bulb was suspended above the desk giving a rather poor illumination, but perhaps that was because the morning sun was growing stronger.

'I don't sleep that much, Mr Monroe, not since my son went into the jungle, never to be found. And that was six years ago. So for some insane reason I still expect him to knock on that door, early morning just the way you arrived, and that makes me a little jumpy. Now what use can I be to you Mr Monroe?' Mr Thorn asked.

'Um, did you happen to receive some news about our arrival..., I mean it should have been sent to you straight from the President's office.' Vincent uttered nervously. 'I am Colonel Monroe and I've been....'

'Now you're talking,' the older man spoke unexpectedly, 'Colonel Vincent Monroe, yes I was made aware of your arrival, it is a boat you're looking for?'

'Yes, and I believe that you've been paid on our behalf to rent us a boat, is that right?'

'No, that's wrong, where are you going Colonel and for what, if don't mind me asking,' the man said, to Vincent's consternation.

'That must have been explained in the telegram, anyway we are here for a scientific expedition,' he quickly answered, still in much bewilderment.

'Scientific huh, what do you know, a military man accompanied by a bunch of mercenaries and a couple of educated people, Oh I saw them alright. The trouble is Colonel; I no longer rent boats, not going upriver anyway, so that's why I said you're wrong.'

'Why is that Mr Thorn?' Vincent dared push the issue.

'Because I never see them back, that's why. You see Colonel, you're not first to come here asking to hire a boat, there were many and they never came back. So that's bad for business. And if you're going upriver with your cowboys, make no mistake Colonel, but that jungle out there, you and your companions will be swallowed by it with no warning,' he said, his voice echoing deeply. 'So I responded to your President and told him unless he wants to buy the boat, there is none. Here is a letter I got in reply, it says that you carry some cash and you'll pay me for it, please read it yourself.' Mr Thorn then explained more casually.

Vincent didn't react at first, but allowed this news to sink in, and then thought that Mr Thorn must be indeed, deeply affected by the disappearance of his son. He took the letter instead and examined it thoroughly and it all seemed to be authentic, carrying the signature of the President. It was sent around the time they set sail into the Atlantic. Vincent nodded and placed the piece of paper back on the desk.

'I am surprised that he did not contact you on the "Elvira", to let you know about our new arrangement.' Mr Thorn commented.

'You see, the crew on the "Elvira" don't know much about the reasons for our presence on the Amazon, they think we're down here for business, I'd like to keep it that way, so the President would not contact us directly.' Vincent explained.

'What is it you're after, Colonel? I mean to have the President backing up this expedition; it's got to be a hell of a thing.'

'That it is but I can't say no more, shall we get down to business? How much is your asking price?'

'For the sort you looking for, selling price, well I think ten grand is fair.' Mr Thorn said circumspectly.

'I tell you what Mr Thorn; I will give you seven, all cash naturally. If we bring that boat back, we get five thousand from you and so you get the rental fee for it, three months maybe four, I guarantee you'll have your ship back, how is that?' Vincent requested confidently.

'Well Colonel Monroe, this is not the States, this is the Amazon and I hardly negotiate with anyone, even on behalf of the US President, he's got no jurisdiction here, get me?' Mr Thorn replied harshly.

'Have a good day Mr Thorn, apologies for my interruption.' Vincent said as he got up and saluted the old man by picking up his hat, which he sat back on his head.

'Hold on, seven it is, but if you bring it back I only give four thousand in return. Like I said this is the Amazon where you won't get a better deal, damn it,' Mr Thorn cursed through his teeth.

'Done deal sir.' Vincent said turning around to face the peculiar Mr Thorn, 'Now I am going to need some supplies and please tell us where we can crash till the morning before setting course upriver,' he enquired feeling good for not budging much on the bargain.

'I will fill it up with fresh drinking water, some food and fishing rods. Don't drink water from the river unless you boil it first, it will kill you after a long sickness I am told. Please sit down Colonel, I have a few more hints to give you,' Mr Thorn pleaded.

'Is that what I think it is,' Vincent asked, his attention being focused on an object placed at the opposite corner, upon a tall table, only visible as he turned towards the door. 'It's the real thing, isn't it? Haven't seen one this close,' he added as he sat back on the chair with his body half turned backwards still admiring the odd piece.

'Thomas Edison's home phonograph, two minute cylinder, I use it when I feel lonely..., and that's quite a lot of times.' Mr Thorn explained with an air of sadness. 'Out there in the forest Colonel, I am told to stay away from everything that crawls, walks, flies, swims or even stands, you get me? It is full of beasts, God only knows, I only seen monkeys, snakes and some jaguars, some of them were caged. You see there are many people coming to catch these mindless monsters for showing 'em up around the world, good business I learnt. Then there is rubber they're after perhaps not so much anymore, but mostly people come here seeking treasures, for Pete's sakes, I just don't know how they think to find it even if it were hidden out there, El Dorado they say, the city of gold, how sad..., never seen 'em coming back. Is that what you're after Colonel? Are you seeking El Dorado too,' he asked after his elaborate little speech.

'No sir, haven't even heard of it.' Vincent replied calmly.

'Um not that it is my business and I respect what you told me earlier, we'll keep it quiet, but take it from me, unless you're well up to it that jungle will see the better of you as it's seen the better of many before you, including my own son, damn fool, he got everything one will ever wish for, but huh, sense of adventure, so much of it that got him killed, I think,' Mr Thorn explained, the sadness emerging again in his tone.

'I am sorry for your loss, sincerely sir. We'll be careful and always take account of your advice.' Vincent told him as he stood up once again. He nodded and walked across to the door. 'I will come back with your payment, if you show me the boat,' he articulated intending to open the door.

'Yeah, yeah, the boat, it is the last one, farther up the pier, "Mary Rose", I named her in remembrance of my wife. It is the last one in good shape at this time, that's why I was asking for that kind of money, it's a good boat Colonel.' Mr Thorn raised his tone.

'I'll be back in a little while with the money and we'll take good care of it. You'll get it back, I can assure you Mr Thorn,' Vincent said as he disappeared through the door.

Once outside, Vincent signalled his men to follow him, and then they walked quietly to the end of the pier. "Mary Rose" was anchored just there and it didn't have much of an inviting look, it appeared to be a rusted piece of junk, lying solitarily, many yards away from any other vessels. It was a medium-sized square platform that formed into a triangular shape at the bow, on which one deck galley, storage and perhaps small cabins for the crew were placed, while on top of it there was the forecabin. It wasn't much different from boats seen on the Mississippi or Colorado rivers at the time, but obviously modified for it had no large paddlewheels set astern, neither on the sides. It had just one horn set right behind the forecabin while the stern held a flagstaff with a Brazilian flag hanging above the American one. The deck and cabin were supposed to be yellow that was a dirty yellow now, and the rest of the hull appeared to have been blue but looked more like black while rust patches were observed everywhere on the deck and the cabin. All in all "Mary Rose" looked old and exhausted perhaps due to many years in service; surely it had some stories to tell if it could only speak. It wasn't attractive and didn't feel that safe, but for Vincent it inspired personality maybe based on the ship's long history. He liked it and in spite of the price he felt rather comfortable taking it. Choices were limited and the expedition must go ahead, no matter the boat or the cost of it as long as it sailed.

'Gentlemen this is your new home for the next while,' he said rather approvingly, to the pure bewilderment of all the others.

'Boss, have you lost your mind, if we step on the damn thing it will sink before you start the engine!' Wade said in disbelief.

'Blind me!' Mr Englewood added.

'What is it gentlemen, did you think I walked here to look at this boat just for my own amusement.' Vincent replied with an ease of arrogance, 'This is it, get things aboard while I pay the man and then we can go to see the town, get some good sleep someplace and we sail at first light tomorrow,' he added, suggesting there was no room for arguments.

Without waiting for any more complaints Vincent walked away, after taking one more glance at the boat. Sincerely, he thought, it wasn't worth that money but he felt that in spite of its looks, this boat was running well. He decided to return to see Mr Thorn, perhaps try to negotiate a little more on the price, recognising that he rushed with the offer before seeing the merchandise. Even so he still thought that he did good bringing down the asking price. As he walked back inside the office, he found Mr Thorn writing something in his register and therefore kept quiet allowing the man finish his work.

'Oh sit down Colonel, just getting ready with your paper work, how do you find the boat?' Mr Thorn said, his head still down at the register.

'Doesn't look the one million dollar piece now, so I was thinking we might discuss the price a little more. I know I gave you an offer and as a gentleman I won't take that back, but maybe you can throw us some fuel for free.' Vincent said informally.

'Um, that figures, I guess, you can get one hundred gallons of diesel free of charge. You see, my son was a brilliant engineer, so he converted "Mary Rose" from a steam engine to be powered by diesel. He also got rid of the paddlewheels, replacing them with modern propellers made for shallow waters. So when you look at all these, the price I am asking is not over board, so to speak, besides it has sentimental value attached to it.' Mr Thorn explained gravely.

'I can understand that Mr Thorn, I will be happy to give you the amount agreed upon and sign your register.'

'Do you mind my asking but do you have anybody capable to sail the damn thing?'

'Sure, Mr Wade Johnson out there, he used to be a naval officer, back some while ago alright, but very skilled as I can recall.' Vincent replied confidently.

'Naval officer says you? Military then, tell him this, he needs to keep the boat as far as possible from the shores, don't get near unless you must. Even with all the adjustments the boat can still get stuck, but more so beware, down here Colonel the enemy strikes first and then if you survive, you might get a chance to see it, but not for long for it strikes again and again until it makes sure there is nothing left. Oh, I heard stories and no matter how incredible they sound I think they are real. I can show you so many entries in my register here, brave explorers they claimed, well acquainted with danger but once they went, they were never heard of again, amongst them my adventurous son. I don't want to scribble a cross sign just beside your name, do you hear me?'

'Loud and clear Mr Thorn, I promised I'll return that boat to you and I will do just that, hopefully in just a few of months.' Vincent added with conviction.

'So help you God, Colonel Monroe, so help you God!' Mr Thorne said, his voice resonating enigmatically.

A few hours later Vincent found himself lying down on a comfortable divan inside a room at the "Tropical" hotel. It was nice and clean, quite a new luxurious building, right at the heart of the town. The price was proportionately high, but a few hours in comfort after weeks on the "Elvira" and before God knows how long they would spend on the uninviting "Mary Rose", Vincent thought of it as a well-earned treat for all of them, designed to increase morale amongst his mates. Most of the other lads had quickly hit the beer garden as soon as they walked into the hotel, but Vincent decided to rest, something that he strongly advised his companions to do as well, but to no avail. Drinks, women and amusement is what they had in mind and perhaps they were right for there was a hard journey ahead. It was a journey that had nothing but hazard written all over it, a journey that might well mean the end of their lives, if not for all of them at least for some of them. And if so far Vincent hadn't give much thought to the seriousness of the situation, after listening to Mr Thorn, he realised how poor was the judgement made when he agreed to come all the way here in search of "God". Perhaps this very last day of comfort would mark the beginning of an endless struggle in which the purpose itself became secondary to keeping alive, which now became the most important goal. Ultimately it was down to their skills, their will to live and their experience, to face and deal with deadly situations, qualities which Vincent thought they had plenty of, but there was this other thing; they were about to face danger of a different nature, as explained by Mr Thorn, the enemy is unseen, it strikes again and again until nothing is left and then it goes. In principal they were heading into something much closer to hell and not at all "God" or the "Garden of Eden". Just to see the vastness of what some people call "Green Paradise", it would be impossible not to think of it as the "Green Hell", stretching to all four corners of the world.

Chapter Six

The noise coming out the "Mary Rose" diesel engine sounded rather indolent, almost like a melancholic moan although it was mechanical in nature. The boat was gliding with ease, upriver after taking a right turn back onto the Amazon. It was almost dark as the crew delayed the departure by several hours. However it was all good, people in high spirits, plenty of supplies aboard and a rather positive sense of anticipation was floating in the air. People were laughing and making bad jokes about one another after spending much of their time amongst the whores in the hotel, soaking their minds in exotic local beverages. But now they had passed that moment of indulgence in spite of some being still stuck in the euphoria. As for the three men of books with Vincent amongst them, they were more focused on planning the days ahead, days that could potentially change their good moods into sorrow, something not desired by any one aboard.

'We have still hundreds of miles to cover by water and if that gives us some sense of security, most of the journey thereafter is on foot gentlemen.' Vincent explained looking down on the maps given by good old Mr Thorn.

'Yeah, that's when we're going to face some real danger.' Professor Mathews added.

'According to some of the indigenous people these waters are not very safe either.' Mr Englewood persisted.

'We'll try to sail right through the middle for as long as possible, the waters are deep for a good bit still and then let's hope that Mr Wade knows what he is doing. Here we follow the Amazon River or the Rio Solimões as the Brazilians call it, all the way up to the confluence with the Ucayali River. How long do you reckon Mr Wade?' Vincent shouted up to Wade Johnson in the front cabin.

'At current speed, gentlemen, if we don't stop too often I'd say ten to twelve days, but these motors need a break now and then. Maybe fourteen days, then add about four more before we step down from this damn boat. We'll use most of our fuel by then, but downstream should be a smooth ride on our return.' Wade shouted back to them.

'So it's roughly three weeks before we can walk again. We'd be nearly at the edge of the forest at that point, close to these Andean plateaus described by Mr Rasmussen in his notebook. We'll use his maps to orientate ourselves and find the place, please God they are accurate. The journey on foot then will take up perhaps seven to eight more days to its endpoint.' Vincent concluded.

'...A hell of trip then gentlemen, a hell of a trip.' Dr Jacobson added.

The dusk quickly turned into darkness while a much cooler and less humid air begun to set in. They were also quickly leaving civilisation behind them with every mile "Mary Rose" was pushing upstream. It was already hard to imagine civilisation whilst seeing this imposing wilderness all around and in spite of only having left the port only a short while ago for a matter of hours, the idea of humanity was shortly to transform into some distant memory, because most of this vast rainforest was hardly touched by it. With the ever growing darkness, countless sounds as if from hell began to echo on both sides of the boat as the children of the jungle were coming alive. Some sounded far in the distance; some others were close, so close it felt as if they were right there aboard the "Mary Rose". It was harder and harder for most of the people aboard the ship to stay relaxed, and most of these men had been weathered by hardship, by war and by killings, but in front of this great wild and unfamiliar environment combined with the noises from the inferno, previous experience and hardship would hardly matter.

'Is it God we're going to search for boss, cause I got to tell ya, those must be his pets howling out there...' Running Wild commented sardonically.

'How can you be so placid about it?' Mr Englewood enquired, his voice very tense.

'If I was any other way, what purpose would that serve Professor?' the half Indian replied.

'Yeah I heard you people have nerves made out of steel,' the English man said in return.

'They're nothing but myths, Mr Englewood, and I ain't even purebred, my father was Irish, just my mother was Hopi. Nothing inside me is made out of steel, while composure, well that's something you acquire after many tests, each time you survive yet to live another day, but the scars are always there to remind you about the times when you're edgy, rushed and indecisive. So yes Professor I am placid cause I've learnt from my past.' Running Wild concluded in the same calm manner.

'Huh, what do you know, half native, half Irish and a philosopher on top, that's some bloody combination isn't it?' Mr Englewood said ironically. At that very moment he felt a strong hand on his right shoulder, pressing it hard. At the same instant, he heard the repulsive sound of spitting as a big hunk of saliva landed in between his shoes.

'Listen Professor, with all due respect, we don't appreciate your sarcasm around here.' Clive Orson spoke gravely in his ear just about releasing the pressure on the English man's shoulder.

'That's enough gentlemen, save your breath, you're gonna need it for later I'm sure.' Vincent prompted the two men.

'Nothing but a bunch of bloody cowboys,' the professor cursed, leaving the deck.

Days and nights alternated on their journey upstream without anything significant taking place aboard the "Mary Rose". However the living conditions, under extreme heat and humidity, movement restrictions and so on, allowed fatigue to impact severely on some of the people. They were getting sick one after another experiencing nausea, high fever and the likes, but thankfully Dr Jacobson was well prepared and capable of helping people get over these symptoms before the sickness developed out of control. It was day six since they left the port town of Manaus and the boat made considerable progress up the Amazon River. In spite of being aware that they shouldn't get close to the shore where waters were shallow and filled with dangers, some of the crew suggested that they should. A break from the deck, a bit of freedom of movement and perhaps something else to feed upon, other than fish and canned food would be more than welcome, they said. Vincent agreed eventually, although he was strict and unlikely to give in to pressure, he thought, in spite of the risks, people might feel better afterwards and continue to be motivated for the rest of the journey.

'Extreme caution gentlemen, approach the forest with extreme caution. We'll deploy the rescue boat and only four of us at the one time will step on land. We shall look for a place where the vegetation is not very abundant for we are less likely to encounter alligators and the sort. If we're lucky enough to catch some bird or other edible animal, fair enough, but I won't make that a priority. It is risky and time consuming,' he explained to his companions.

'We hear you, boss, but just four of us at a time will mean a couple of trips and that will also take time.' Clive replied.

'One hour tops for each crew, if that's enough and nothing goes wrong, please God, we should be on our way before sunset.' Vincent concluded.

'Sounds fair to me,' Running Wild commented.

'Yeah Wild, you'll be one of the first to go, based upon your experience in dealing with the wilderness, don't you reckon?'

'I appreciate your vote of confidence boss.'

A little while later the first five people were rowing away towards an area that looked more like a beach, which had a significant portion of it, bush free. They were five because it was agreed that Dr Jacobson should accompany both trips in case of unexpected injuries that required immediate attention. The five men where feeling a little apprehensive as they approached the shore on their right hand side. The tree line was at only about fifty yards from the edge of the beach and no one could tell what sorts of beasts, if any, were watching their arrival. Surely they needed to be on their toes, always ready to react in case of unexpected encounters. On the other hand considering the vastness of the jungle it would only be by bad luck that they would come face to face with unwanted company. Running Wild jumped out of the boat first and rushed to the stern, pushing it hard so that about half of its length landed on the sandy beach. The other four men stepped out moments later and pulled the boat up its whole length so that it was secured on the beach. Running Wild walked towards the tree line after signalling his mates to wait quietly until he indicated otherwise. He then spent several moments in total silence right at the tree line, as if inspecting the place thoroughly using his senses to their ultimate level. Minutes later he returned to the group appearing a little worried but not totally alarmed.

'I can't just be sure; this place is different to anything I've seen before. We may be safe for a while but not too long, I just can't tell,' he explained.

'You have no fear Wild, but now I see you're concerned. We'll have that in mind. Folks if you have anything useful to do I say you do it now, like try and move around the place and stretch your bones. I and Wild will try finding something worthy of a shot; get some fresh meat for dinner tonight. Mr Englewood, Mr Jacobson, you stay close to Mr Orson there, do as he says and keep it quiet all the same,' Mr Wade told the group before turning towards Running, 'did you see anything that could be classified as dinner?'

'There's a lot of action in those trees but I think they are just monkeys.' Running told him.

'Never tried one before and from what I heard being said back in Manaus, indigenous people eat 'em alright.' Wade said.

'Are you complete savages, they are like our distant cousins for God's sake, it would be like eating human flesh almost.' Mr Englewood protested in pure bewilderment.

'Almost but not quite is good enough for me and by the way I was kidding Professor, and that thing you said about distant cousins, I can tell you, I got nothing in common with monkeys, don't know about you.' Wade replied in a more serious voice.

'If you ever heard of a man called Charles Darwin you'd perhaps change your views,' the Englishman whispered.

'Yeah we've heard of your scientist and his theories on evolution, but they're just theories, am I wrong?' Running Wild commented confidently.

'You never stop surprising me, Indian,' the professor said almost in shock. He had to admit that the cowboys and the half Indian were not as stupid as he presumed at the start of their voyage. On the contrary they seemed articulate and it now seemed likely that at least Running Wild had benefited from either formal education or a good insight into general knowledge. And for that matter he wasn't the only one as their leader, Colonel Monroe appeared well in tune with matters of that nature as well. Professor Englewood felt a little embarrassed by his conduct so far and decided not to treat any of these people with any air of superiority. He nodded in a half-hearted sign of apology and stepped to the side only to nearly bump into Clive Orson who displayed his teeth – which were not at all well looked after and rather disgustingly dirty - in a smirk that betrayed his dislike for the Englishman. Mr Englewood felt that he would rather commit suicide at this stage than have this monstrous cowboy babysitting him for the next hour or so.

'Okay gentlemen, this debate is useless, we'll try and hunt some bird or even better if we come across some boar, God only knows the odds.' Wade said as he walked off to the tree line while loading his rifle. Immediately Running Wild followed in his footsteps leaving the other three men at the boat site.

'Well now, you two ladies can walk up and down the beach but always in my sight, we clear?' Clive told them in an arrogant tone to which the professor decided not to reply.

For over forty minutes, both the doctor and the professor ran slowly up and down and performed stretching exercises, to the amusement of Mr Orson.

'I wouldn't have missed this for the love of God,' he laughed to himself.

As he continued with his laughter, getting even louder now, a thundering noise split the air and then another, as two gun shots disturbed the silent atmosphere dominating the picturesque surroundings so far. In that instant several birds took off from the top of the trees making a deafening noise, which combined with piercing screams of monkeys and perhaps other beasts transformed the whole scene into an inferno.

'Hell yeah, that's the jack pot baby!' Clive shouted as he began running towards the tree line.

After few moments in which the doctor and the Englishman got over the confusion, they too rushed up to the tree line. All three were enthusiastic as they stood there waiting for Mr Wade and Running to emerge from the dense vegetation with hopefully a big prize, maybe a pheasant of some sort or a boar, anything that could be classified as a tasty meal full of good proteins. Some time passed by and nothing was happening as Mr Orson went down on his knees to try to peer the thick vegetation for any sign of the approaching men.

'Do you think they got something?' Dr Jacobson inquired almost fearfully.

'Yeah they got something, those two boys wouldn't waste a bullet for no reason, I can assure you,' Clive said very calm still.

'I hope it's not monkeys…,' the Englishman whispered.

Mr Orson just shook his head in disgust but had no time to respond to the professor's comment, as the bushes in front of him began to bend to reveal his buddies emerging from the thickness of the jungle.

'That's a score gentlemen, two boars, mother and baby it seems,' Wade Johnson spoke as he came out in the open and threw a medium sized piglet at Clive's feet, 'here we'll give a hand to Running, he's got the big one,' he added.

Moments later a bigger animal was laid down beside the other as Running Wild dropped it from his shoulders. The wild pigs did not seem awfully big, not even the adult one, perhaps weighing around sixty pounds. They were covered in grey fur and each had a big red spot on the neck, where the bullets had penetrated their thick skin.

'Unusual looking, I must say, I don't think they are known to science,' Mr Englewood commented after a close inspection of the two mammals.

'They'll be known to your guts Professor, that's for sure,' Wade laughed. 'Let's get them on the boat, I'll start skinning the bastards right away,' he suggested.

'Not a chance Wade will do them here. I think it's better not to leave trails of blood, God knows, it may attract alligators chasing our boat.' Running Wild concluded.

'That's clever, maybe Vincent and Ramon can do the job; let's just pull them near the bank then.' Wade said.

Not long afterwards the two animals were laid just beside the boat. Running Wild took a couple of handfuls of sand and rubbed it against the shot gun wounds on the animals, attempting to eliminate as far as possible the smell of blood dispersing in the air.

'I think the doctor should stay behind, the others will be here like in minutes, what do ya think?' Wade Johnson proposed.

'I'll stay with him, I can start skinning the pigs, I am quick at it,' Running Wild said casually.

'Good idea, let's go people, we're well over the hour mark,' Clive Orson shouted as he pushed the boat back into the river.

As the boat returned with the other three men aboard, Running Wild was almost done skinning and gutting the young piglet, after he quickly dug two holes in the moist ground in order to throw in the skin and the guts and then cover them with sand.

'Cleverly done Wild, here I will help with the other one. Gentlemen if you could just take your time stretching your legs, but not going too far, we'll be on our way shortly. I am sorry Mr Mathews but I don't think we'll spend the whole hour here this time, too dangerous,' Vincent said politely.

'Fair enough,' the American professor replied.

They were almost done as Vincent and Running were now loading the skinned animals onto the boat, being laid on a sheet of rubber to prevent any blood dripping in the water. The other two men were on the far side, on the south side of the beach, apart from each other as Dr Jacobson was ahead as he returned to the boat. Professor Mathews was walking slowly behind within a few yards of the doctor. Running Wild turned towards them shading his forehead with his right hand in order to protect his vision from the blinding passed midday sun. Without saying a word he quickly grabbed his rifle which was leaning against the boat and began to rush towards the two men. As he reached the halfway point he went down in one knee and secured the rifle on his right shoulder aiming straight at the professor. In pure bewilderment Mr Mathews froze right there in the middle of the beach, totally unaware of the threat coming from behind him. A big yellow cat with brown spots was moving in slow motion, its body tense ready for the assault.

'Don't move Mr Mathews, don't move,' the half Indian man whispered slowly pointing the gun towards the cat, trying to get a clear shot. Many yards behind him Vincent was standing with his rifle aimed in the same direction; only he was way out of range to have any chance of hitting the animal. In that moment the "expected unexpected" happened as Mr Mathews attempted to turn around after realising that he wasn't the target of the two men, but that something else must have been behind him. His sudden move proved fatal as the big cat lifted its body into the air with ease, embracing the man's body with its front limbs, landing its powerful jaws at the junction between the professor's neck and his right shoulder, penetrating his flesh with such power that the man's body collapsed instantaneously. The big cat suddenly began clawing at the professor's chest inflicting more deadly wounds to his thoracic area. Then it rapidly started to pull the inert body backwards into the bushes nearby, effortlessly in spite of the man's weight. Running Wild squeezed his rifle's trigger with a booming noise but missed the target completely. The only effect his gun shot had was to make the feline abandon the body for a short moment as it tried to run into the bushes, but it quickly returned and grabbed the body again, attempting to pull it all the way in. Thunder after thunder pierced the air as Vincent launched his assault, firing his rifle while running towards the yellow cat however not aiming precisely, his bullets were only impacting on the sand nearby. With a deep snarl alternating with high snivelling noises, the big cat ran in circles for a few moments before disappearing into the deep verdure, showing no more interest in carrying the professor's body with it. The three men ran to Mr Mathews' body which was now jerking violently as he went into convulsions. The blood was pouring abundantly from the base of the right side of his neck and also from the chest area. Dr Jacobson pressed hard on the neck wound shouting to the other two to try and cover the chest wounds in an attempt to stop the heavy bleeding.

'If we cannot control the haemorrhage, we won't save him,' he articulated almost out of breath. 'What the fuck was that?' he asked.

'A jaguar, a goddam jaguar, he must have smelt the blood from the pigs, I'm sorry, I should have known it was too risky,' the half Indian man explained.

'It's not your fault Running, it isn't, is there anything we can do to save him doctor?' Vincent said.

'I am afraid not much, not without proper medical care like in a clinic, he is losing too much blood.' Dr Jacobson replied while tearing down his shirt's sleeves trying to make a bandage of some sort. 'Here press hard on the neck wound and then remove your hand when I finish wrapping this cloth,' he said looking straight at Vincent. One piece of material which the doctor folded in a square shape was then handed to Vincent in order to press it against the wound. 'Now lift his head slowly so I can wrap this,' the doctor indicated. But as he spoke the professor opened his eyes, giving Vincent a desperate look as if asking for help. He then grabbed his arm trying to lift himself up while moaning something hard to make out.

'Not now, Professor, try and stay calm, don't speak…' Dr Jacobson attempted to say, when he noticed a massive outpouring of blood coming from the man's mouth. The doctor shook his head as he listened to the professor's deep coughs, a sign that he could do nothing more to help him. The man's body started to shake again viciously as the coughs increased both in frequency as well as intensity. Vincent looked up at Running Wild and nodded before getting up and stepping back. In that second Dr Jacobson threw his weight backwards as well, as Running's rifle thundered once more making a considerable hole right in the middle of the professor's forehead. The jerking stopped instantly and so did the coughing, while the blood was now coming out in huge amounts at the back of his head.

'He's done, we can't bring his body onto the boat but I can quickly bury him here.' Wild said remorsefully.

'Is there any point, I mean the beast will uncover him anyway, and why waste more time in harm's way?' Vincent enquired.

'You're right, we need to go, we need to go now.'

'We're not leaving him here, for the love of God gentlemen!' Dr Jacobson shouted in sheer panic.

'He is gone, sorry Doc, but he is gone. If it's any comfort to you there is no more pain, but we can't carry him onto the boat.' Vincent explained gently.

The evening rapidly progressed into night. There was only one noise filling the air all around, the mechanical beat of the "Mary Rose" diesel engine. Everybody was silent aboard the vessel, deep in thought, unable to put their thoughts into words. It was a tragedy for them all to have lost a man the minute they stepped onto dry land, something that was raising serious questions. Should they just turn back? Was there any chance to actually complete this expedition at all, not to mention without losing more lives? Nobody could answer and Vincent wasn't even entertaining the idea of turning around. He was now confined in his small quarters while all others were still on the deck. Clive Orson was cooking some of the boar meat in spite of comments thrown at him by some of his companions.

'We have to eat,' he told them, after pouring a serious amount of whiskey down his throat.

He was right perhaps, what else? The journey was still going ahead and if they were to complete it, they had to stay strong, including physically strong, as for their morale, well that was already damaged, perhaps beyond repair, but with no alternative they had to move on. But moving on also meant facing yet more dangerous situations, mostly unexpected encounters with ferocious beasts that would not hesitate to turn humans into much desired meals for themselves. For that matter they had been warned back in Manaus by more people than just Mr Thorn that attempting to go far into this jungle may well mean no return, at least not for all of them. Dr Jacobson also speculated that he was limited in terms of tools and medicines and therefore hardly able to save their lives if need be. Besides acquiring injuries from encounters with wild beasts, there were all sorts of diseases waiting for them, mainly due to poor living conditions and lack of proper sanitation. Over all, this entire adventure had nothing but bad prospects ahead for it, which these few men were hardly going to anticipate, never mind to overcome. Vincent spent the whole night thinking about this, finding it difficult to sleep. He couldn't face going back empty handed, but these were his thoughts and feelings, and perhaps he should ask the others if they wanted to continue this dreadful journey. He figured that if they wanted to return the time would come for them to do that after they reached the drop off point. Somehow he felt that they wouldn't, but he also believed that he owed them the opportunity to choose for themselves. Nevertheless he wasn't going to ask them now, because he needed the boat to go all the way until he could continue the remaining of the journey on foot. After many hours of struggling between agony and hope he decided to get some rest for he was no good to anyone without it. But he did not settle before making another entry into his journal. The next morning his diary read:

It is June 6th 1912; we're now nearly a week after leaving the port of Manaus. After a few days that went by rather peacefully, we have confronted the real menace that this jungle has set for the weak and the ignorant. We always felt that there was something watching us gliding silently through these waters, but which could not have touched us for we were well away from it. However this morning we decided to step off the boat as it's like being imprisoned on the deck, with only a few yards to spare to move about. And so we did find a beach that felt safe, well safer than other places we've seen before. It measured a couple of hundred square yards in total perhaps, and the open space gave us a sense of security in so far as we thought we wouldn't be taken by surprise by anything. Even if we were, we would have time to react and push danger away, but we were wrong. We were so wrong in fact that Professor Alan Mathews has paid the price of our ignorance with his own life. It happened because Mr McNeely and Mr Johnson hunted boars in the forest and then we skinned them on the river bank, letting the smell of blood travel some distance into the forest. It wasn't long until this jaguar viciously attacked Mr Mathews from behind, leaving us no chance to save him. It happened quickly and unexpectedly so that I and Mr McNeely could not even take a clear shot at the ferocious cat. We did fire our rifles but were barely close to the target. However we did manage to scare the beast off as it ran back into its hiding place. We and Dr Jacobson attempted to save the poor man's life but to no avail, not without proper medical tools. Unfortunately we couldn't do much for the badly injured Mr Mathews. He died after a few minutes struggle and as he was apparently in a lot of pain "Running" decided to inflict a deadly wound and put Mr Mathews out of misery. I blame no one for this tragic misfortune that caused Professor Mathews' death, but I have to admit to our ignorance of Mother Nature, and this jungle, which is far greater and far more unpredictable than we could have imagined. I hereby promise to personally apologise to Mr Mathews' family on my return to the States if lucky enough to stay alive. Signed: Colonel Vincent Monroe

Chapter Seven

This rain was pouring for days now, with some intervals of sunshine, but overall it became very frustrating, as on a boat the choices are one's cramped cabin or the rainy deck, neither of which was appealing to the men aboard the "Mary Rose". Moreover Mr Gonzales fell ill with malaria but luckily the treatment with quinine based anti-malarial medicines, brought by Dr Jacobson, had produced much improvement in his condition. The other crew members weren't in the best shape either, their physical health was beginning to suffer due to lack of exercise. And then psychologically there was an accumulation of factors that made most of the men suffer some depression, often becoming irascible with one another and have anxieties about the future of this journey. The rainforest had already claimed the life of one man, made another one critically ill and even though he was stable it wasn't pleasant to see the symptoms of it, which were rather scary and left the others pretty fearful of what was to happen next. The boat was almost at the turn onto the Ucayali River at the confluence with the Maranon River and once on it, there were only few more days remaining until they would finish the river journey. But before that a little reward was waiting, a couple of days stop in the Peruvian town of Iquitos At this point in time they were all in agreement about continuing the journey to the end, after Vincent had given people the opportunity to choose if they wished to return. He proposed that if they wanted to stay in Iquitos, they could return to it after dropping Vincent off. However they had to promise that they would go back to meet him at the drop off point after four weeks, wait for a few days at least and then hopefully all of them would see their way back to the States. They all turned down the idea without much consideration, claiming that perhaps more awareness and fewer hazardous actions might be more beneficial than facing useless journeys up and down the river if the group were to split up. Vincent was pleased to see the determination in the men and their response gave him a sense of confidence, something which he hadn't felt when they set off for this journey. But for this resilience to last he needed certain things to happen, certain things not to happen and all this combined with a fair amount of luck. And Vincent knew well that in this equation, with way too many unknown entities it would be virtually impossible not to come face to face with danger and even more deaths among his crew. However for some insane reason he was very determined to complete the journey, collect the data and return all the results back to the States, no matter what his companions might decide for themselves. For now though the stop off point, in a place claimed by human civilisation in the heart of the wilderness, was not only desired but much needed and anticipation was building up with every mile they covered. The sun above was showing its face again after many days of prolonged rain and it appeared to bring up the spirits aboard the "Mary Rose".

'Fucking hell man I thought that rain would never end.' Wade Johnson commented.

'That's why it's called "rainforest", you twat,' the Englishman whistled through his teeth.

'What did you say?' Wade turned around in anger.

'Nothing, really was nothing, I just said that it rains because we are in the middle of the rainforest.' Mr Englewood explained a bit breathless.

'I know where we are professor, but that doesn't mean one cannot say it's enough when it is enough,' the skipper replied as he climbed up into the forecabin. 'Look,' he shouted 'there is a boat on the river bank to the left!' Wade continued enthusiastically, sticking his head out the front window.

'...anybody aboard Mr Johnson?' Vincent enquired bringing his right hand to his forehead.

'Not that I can see boss, here catch..,' the skipper said dropping down a telescope.

Vincent took a fair bit of time to scrutinise the abandoned vessel as they were getting closer to it by the minute. There was no one in sight on the deck, nor on the ground nearby. It rather looked like the boat had crashed into the shore as a big part of the elongated bow was caught on the narrow sand strip, in between some bushy vegetation. As "Mary Rose" got closer the picture became clearer for those on board that this boat had landed there, not by a deliberate action but as the result of some accident. Several marks of a heavy impact were seen on the port side, while the mast looked as if someone had chopped it with an axe right down the middle.

'Do you see what I see, skipper?' Vincent shouted, not leaving the telescope yet as he wanted to see if there were signs of life.

'Yeah, I see it alright, that boat has gone through some rough times boss.' Wade replied. A few of the men gathered around Vincent at the same time, watching the terrifying wreckage up ahead.

'Should we get close to it?' Wade inquired loudly from his cabin.

'I don't know Wade, what you reckon, should we just ignore it?' Vincent replied sounding very unsure.

'What if someone needs help?' Running Wild asked.

'I just can't tell Running, hey skipper what do you think happened here, did it run into the shore?'

'It looks as if it's been through a storm boss, but that would be more likely if it were at sea, not here!' Mr Wade shouted back as he reduced the speed of "Mary Rose".

'That's no storm that did that,' Running commented in a low voice 'it appears that it has been smashed all over and there is nothing to crash into on this river. They lost control of it but due to something else, not a storm,' he said raising his voice.

'What do you say then, is it worth having a closer look?' Vincent asked him.

'I think so, but you're the boss,' the half Indian said.

'Wade, we'll anchor the vessel, try and get close to the site in the small boat. I will go with you and Running, have a peep and get back.' Vincent explained.

'Will do boss,' the skipper replied bringing the vessel to a halt. In an instant Clive Orson threw the anchor down on the river bed. The waters here appeared to be more shallow and narrow. Within minutes the three men hopped onto the small boat and begun rowing towards the abandoned vessel. It looked severely damaged as if some unknown power had come up against it, for the crew seemed to have had no way of saving their boat. But where were they? The whole scene looked completely desolated.

'Hey, is there anybody aboard?' Vincent shouted at the top of his voice. A few minutes passed as he repeated those words numerous times, but they got no reply.

'On second thoughts we probably shouldn't get near this thing,' the half Indian suggested in a low tone.

'What would you say happened here Running?' Vincent asked inquisitively.

'Hard to tell but my feeling is that they came against some sort of a monster. People back in Manaus talked about giant serpents and the sort, you know?'

'What about alligators, could they do that?' Wade inquired curiously.

'Nope, I can't see them able to break the mast, but a serpent powerful enough could…'

'How about the people, you think they might still be around,' Vincent asked quietly 'Hey! Can anybody hear me,' he shouted again. Several noises like screams at high intensity came back from deep within the woods but none sounded to be human.

'I'd say they're lying right down at the bottom boss.' Wade commented placidly.

'You may be right Wade, I just hope not.' Vincent replied.

A little while later Running and Vincent jumped aboard the vessel that was named "Mercy" as inscribed on the port side. This sounded quite ironic because whatever happened here left no room for mercy. On the contrary just seeing the scale of the devastation on the deck, where all sorts of objects were lying in total chaos mixed with several blood trails, indicated a cruel and vicious struggle, the only question being; a struggle against what?

'Any chance that people escaped on foot, went deep in the forest?' Vincent asked realising just how lost he was after seeing the devastation.

'That might have been a possibility boss, only there are no marks on the sand below, so I doubt that,' the half native said. 'Boss we better get the hell back without getting into further exploration, judging by the blood marks, which are still fresh this happened no more than a day ago. Also the coal in the engine hob is still warm. We might be in harm's way,' he added.

'Fair enough, let's get the fuck away before it's too late…' Vincent attempted to finish his sentence only for Wade Johnson to start shouting at the top of his voice.

'There is something big, gotten into the water heading for "Mary Rose", hurry up folks!'

'What is it Wade, did you see it?' Vincent shouted back looking down at Wade.

'No, but look,' he said pointing out towards the river where undulating waves could be seen in an upside down "V" shape, for a considerable length. There was something in that river, something about 20 feet long and moving fast towards the "Mary Rose". Instantly, Vincent began shouting and waving his hands in an attempt to signal to the men on deck, that there was potential danger, but to no avail. They were way out and the wind was blowing in the exact opposite direction for his desperate screams to reach the "Mary Rose". Also in that instant a gunshot pierced the air, echoing in the far distance in all directions, making all sorts of beasts nearby shrieks, as Running Wild fired his rifle. Suddenly two figures stood up on the deck of "Mary Rose" shouting out words that could not be understood this far out.

'Start the engines, move the boat!' Vincent shouted back at the top of his voice. Instead of any reply more gun shots penetrated the air, this time coming from the "Mary Rose". Clive Orson was aiming his rifle at the river, firing off bullet after bullet after noticing the approaching beast beneath the dark waters. He was just attempting to reload his weapon when the deck shook violently under his feet making him tumble backwards, dropping his bullets on the floor.

'Fuck,' he cursed trying to regain balance while the boat trembled again as if being hit from the opposite side. He ran over to the starboard side taking out his pistol, only to see this serpent's head and body lifting up into the air and then plunging back towards the deck making the boat shake once more on impact. Clive fell back again, absolutely terrified of the snake's sheer size and force. It was massive, measuring several inches in circumference and perhaps several yards in length, deep green in colour and letting out a deep hiss that made the men on board crawl with fear. Without hesitation the beast pushed itself forward, sliding with ease on the deck, instantly attacking Mr Gonzales, who was lying on the deck unable to react. The beast grabbed him by his left shoulder and within moments it dived back into the river on the port side, taking the Mexican with it. Clive got himself up and along with the Englishman ran over to the port side. The scene in the river was unequivocally spectacular and equally frightening as Mr Gonzales' body was gliding at high speed on the surface, unable to even struggle against the sheer force that was pulling him. Mr Orson aimed his pistol towards it but didn't fire realising that the beast was way out of range for a hand gun. On the far side near the wreckage, Vincent and Running along with Wade, were watching powerlessly how the Mexican was pulled towards the riverbank many tens of yards away from their location.

'We can save him boss, we can kill the monster!' Wade kept repeating himself. Vincent remained frozen for several moments as if unable to speak. Instead he looked over to the half Indian who shook his head disapprovingly.

'We need to go, it won't swallow him, but it will come back for more. Look,' Running said pointing out to an area in between dense bushy vegetation where a pile of flesh and bones could be seen. 'it spat its' last victim out in order to attack again and there might be more snakes than just the one we see,' he explained.

'You're right, let's get the fuck away from this place.' Vincent concurred.

'What? I don't believe you two, that's our pal in the jaws of the snake for Christ sake, we can kill the fucker.' Wade yelled in disbelief.

'No we can't,' Running told him as he jumped on the boat, 'we may just have a chance to make our get away before the monster returns.'

Moments later the three men were making promising progress towards the "Mary Rose" which in return was now in motion, trying to close the distance between it and the small boat. Once aboard the "Mary Rose", Vincent and his companions found it very hard to bear the event. They kept quiet for a while, hoping that the monster serpent would not chase them. In this situation the creations of Mother Nature proved to be faster and stronger than man-made machines and even weapons. If this snake could dodge bullets from a "Winchester" 22 calibre rifle and still manage not only to stay alive but continue with its' assault, well that made this monster one terrifying "die hard", for which the few men aboard the "Mary Rose" had no answer. The only option was to getaway while praying that Mr Gonzales and the other corpses would be sufficient to satisfy the serpent's hunger.

'Any damage to the boat,' Vincent inquired after several moments of silence, 'I mean anything that we can see?'

'I think we pushed the motor very hard, is all that I can say for now boss,' Clive replied almost sobbing, 'I am sorry folks, I couldn't save him, that bastard snake was too fast.'

'Aye, it's not your fault Clive, but what was he doing on the deck? Why was Mr Gonzales on the deck?'

'He asked to be taken out from his cabin, wanted some fresh air he said, we helped him out after your departure, that's all.'

'Alright, we set course for the town of Iquitos, we'll see if this expedition is worth our effort anymore, but I can't afford to lose more men and this jungle does not seem to spare one's life, put your thoughts into words as we get there, we'll take a vote.' Vincent concluded.

The next 48 hours felt like an eternity for the remaining crew. They were hardly able to speak to one another, fearful that blame would fall upon them, but realistically who was there to blame? They all knew that signing up for this meant risking their lives; they were all warned back in Manaus that there may be no return and yet they went ahead, even after seeing professor Mathews fall prey to the viciousness of the forest's beast. Then the monster snake proved to be as lethal as the jaguar but, what was worse, after losing two men to these creatures, they couldn't even injure the aggressors, not in any significant way. And that fact would have to be considered very seriously because they were hard men, well used to harm, who spent most of their lives in hardship, killing and injuring many, but here their experience was proving to be totally useless for they could not even devise a viable strategy that would keep them safe. No, not in this place where death came with no warning, where the "enemy" was faceless until too late, and in front of such dangers there is no strategy. All this sacrifice in their quest to find "God" was beyond reason and certainly defeated logic, because this insane idea seemed to be based on a farfetched fairy tale, told by some Scandinavian explorer who was now dead anyway.

On the other hand there was something that they had to ponder within reason, a twenty thousand dollar reward each, it was surely something that no man in their position could ignore; as such an opportunity would never arise again. Perhaps paying more attention to danger, being more aware of how to avoid hazard, although faceless and unpredictable, should somehow preserve those remaining of them so that they could get their hands on the cash. As for the two men who perished so suddenly, may God rest their souls, there was nothing they could have done to save them and therefore no one could be blamed. That thought was now embedded in their heads, with little reservation, but for the moment every one of them was contemplating their options. And there was this other thing for which Vincent had chosen them to go on this journey, they were no cowards, but proud individuals who would not accept defeat, who would emerge victorious any other time, and who would not go back to America with their tails between their legs, because then they would die wondering "what if", something that they had never done before. Yes, losing your mates in such tragic and violent circumstances is demoralising but perhaps not enough reason to give up, to hide or to run away from the responsibility that they had all signed up for. And as the boat kept going forward their spirits were driving forward as well, each finding his own motivation for it. If for the "cowboys" money was a strong and very fulfilling argument, for the Englishman and the doctor there was this insane curiosity to find what they came here looking for, to find this otherworldly place as described by the Scandinavian explorer. Also there was Vincent too, who couldn't entirely define his motivation but he looked at it more like a combination of factors. Money was one thing (more that Gina and her younglings would benefit from it), his chaotic life before signing up proved to be a strong reason, but ultimately it was his rediscovered sense of pride. Just hours before docking in Iquitos port he wrote:

It is hard to even recall how we came to be on this expedition, what sort of thing has driven us all the way to the heart of the Amazon. I personally can't tell, not now when we mourn our two companions who fell in the way of merciless monsters. Losing our two brothers seems a very bad thing, but even worse is the manner in which they died, with no warning and with such a short time between them. It seems that caution alone is not enough to stay alive in these waters, but seriously I don't even know what is needed. I have a moral obligation towards all the remaining men aboard the "Mary Rose", maybe to warn them and even advise them to give up, turn around and continue this journey no more, for they have a duty to their loved ones to return home in one piece. The opportunity will come now, once we dock in the Peruvian Town Port of Iquitos. I will talk to the men and ask them to stay in the port, perhaps after dropping me further upriver, so I will continue this journey for reasons I cannot understand, but I feel for me it is different as I cannot say that I have someone looking forward to seeing me return.

I can't say that they will listen, I don't know if they are scared and want to give up, but surely I need to try before it is too late, because in only a few more days this expedition needs to continue on foot, making our way through the thickness of the forest, facing more unexpected ways of dying. Somehow I expect to face resistance, and if so, I need to face facts and accept people's wills, after all they are free and have free will, at liberty to make their own choice and as a result I can only respect that choice, even if it defies my logic, whereby the sacrifice may be too great for the reward. Signed: Colonel Vincent Monroe

Chapter Eight

It felt good to step on stable ground again, knowing that there was no danger, or at least the assumption was that there wasn't. It was also good to see signs of civilisation again, although the town of Iquitos looked nothing more than a large rural and dirty settlement, and not even close to what the port of Manaus had to offer. However it was a good opportunity to rest in more comfortable conditions, but also to reflect upon the options and draw a conclusion for the future of this expedition, which had so far proved to be rather painful. Heavy at heart still, the men aboard the "Mary Rose" stepped onto the shore, each looking forward to some proper food and a long awaited respite, away from the turbulent Amazonian waters. For Vincent though, the priorities were different as he needed to send some feedback about their situation, straight to the White House. He badly needed to do that, in spite of the agreement he had with the President whereby communication should be limited and secrecy kept at all costs, and therefore only in the case of force majeure should this happen. But if losing two lives wasn't force majeure, then what was? So, the hope was that he would find a telegraph machine that could facilitate sending this message, after all Iquitos was an important commercial point on the Amazon, so progress must have reached this place like all the others, in spite of its' rather poor appearance. He asked his men to go look for accommodation while one of them stayed put in order to guard the ship. He explained to them that he'd go looking for western entrepreneurs in the area so that they could get supplies for a reasonable price, something that they had to do anyway, but omitted to inform them about the telegraph part. Vincent just wanted the men not to alarm themselves, although they were well aware of their situation, but it would be wise perhaps to try to get a reply from the White House before letting them choose whether they'd still go ahead or abandon the expedition, in the hope that it would be called off by the President. It wasn't something that he wanted to happen, not for him surely, as he still wanted to go all the way, even by himself, but he could not ask that of his companions. He took a good walk uptown, making his way through the indigenous crowd, a mixture between native Peruvian Indians, Europeans and even Africans. To his surprise Vincent even passed by a Synagogue, which indicated a strong Jewish presence in the area. Approaching the town centre, the buildings were closer to a modern style standard commercial offices, banks, shops etc. And then in the main square there was the General Post Office. The building revealed a mast connected to wires, right at the top, the very thing Vincent needed to see. He rushed towards it, becoming a bit anxious especially because he did not know what he was going to write in the message, but whatever it was, he needed something very short and concise. As he walked through the door he found two workers, a man who looked rather European and a local women, both in their 40's, husband and wife as Vincent was about to learn. There were no other customers inside the office.

'Buenos días señor,' the man greeted him, adjusting his glasses on his nose. He appeared to be rather short, although seated behind the counter, within a glass kiosk, thin and bald on the top of his head. He was dressed casually with a white short-sleeved shirt over which he had a sleeveless V-neck brown jumper. The women, who just smiled without saying anything, was of short stature as well, dark skinned with very black hair pulled back in a ponytail.

'Hello, Vincent replied timidly, 'do you speak English señor,' he inquired.

'Just enough to get by, I guess, what can I help with,' the worker said in good English but with a strong Spanish accent.

'Oh, good! I need to send a message via telegraph, to America, you do have a telegraph is that right?'

'Yes sir we do, it was just installed here last year and it has seen good use so far. Sometime, very busy, people need communication quick, business you know? Is called progress, isn't it right señor?'

'Yes it is, thank God! Now it is strictly private sir, the message I need sending, but I suppose I can trust you with this confidentiality matter okay?' Vincent asked in a curious tone.

'My wife Carmen, she knows no English, but she can encode the words you write on paper, letter by letter, so no problem,' the worker said confidently. 'My name is Manuel, so if you write your message, she will send it off for you,' he added.

'Oh, sure, I am Vincent; may I borrow a pencil and some paper then?' Vincent said while turning towards the female worker who was wiping the main area of the office. Vincent lifted up his hat and tilted his head forward in a gesture of salute. He then picked up the piece of paper and the pencil handed over by Manuel through the glass window of the kiosk and walked over to the opposite side where a small wooden table and two chairs were placed. In spite of not having formulated the text yet, Vincent began writing at a good pace, having in mind that he needed to be concise but also explain everything. A few minutes later the note read:

"Strictly for President William Taft;

Mr President, although we were in agreement to have no communication as far as possible, I feel that it is my duty to inform you of some rather weighty events and our current status. We have reached the half way point to our destination as we have docked at the port of Iquitos, Peru, just a short time ago. With a deeply saddened heart I need to inform you that Professor Mathews and Reserve Sergeant Ramon Gonzales are no longer amongst the living, as they both fell prey to the wildlife of the place, which I find hard to describe. We're all disheartened by these tragedies and wondering if it is perhaps time to abandon the mission before losing more lives. As discussed before setting course on this expedition, it would be my judgement as to how to approach the future of it at any given moment, and therefore I will take the liberty to release the rest of the men if they so wish. However I will only proceed with such action after I get word back from you. Therefore you need to reply to sender as soon as possible. We will stay in this town for three perhaps four days. Faithfully Yours, Colonel Vincent Monroe.

He then handed over the note to Carmen, who took it carefully and had a quick glance over it. She then said something in Spanish to her husband, now preoccupied with arranging some folders on the shelves situated at the back of the kiosk. Manuel turned around and indicated to Vincent to approach the counter once more.

'How are you going to pay for this señor Vincent,' he then asked.

'I only have American dollars if that's okay in these parts.' Vincent replied coolly.

'That's fine, it will cost you thirty-four dollars and fifty cents, if you will señor,' the worker announced.

'Wow, that's some money for a message, isn't it?' Vincent said in sheer surprise.

'It's a long distance transmission, and your message is not the shortest I've seen, as my wife explained it is quite some piece you wrote. This is going by Morse code, now you wouldn't mind giving me the receiver's address, if you still want to proceed off course.' Manuel explained sincerely.

'Yeah sure, here it is...,' Vincent said pulling out a folded piece of paper from his shirt pocket, which he then gave over to the worker.

'Washington, it must be important then.' Manuel commented after peeking over to the paper note.

'Yes it is, and confidential like I said. Now in case I receive a reply, what happens if it comes through after hours?' Vincent enquired curiously.

'Oh, we live in the building and we connected a speaker to our home, so we're here twenty-four hours a day so to speak. Do you expect your reply soon?' Manuel inquired.

'I don't know really, but I will be in town for a few days. Can I check with you tomorrow,' Vincent explained, '...speaking of which do you know any good place to crash?'

'You travel by yourself señor, or you have company? Right across the square there is a hotel owned by an American, he likes money and usually has rooms, it's not luxurious or anything but it is clean and the hookers are good…,um so I am told.' Manuel replied a little embarrassed.

'That will do, perhaps,' Vincent said appearing to ignore the worker's little gaffe, 'here is the money señor Manuel, you may keep the change. I will check in tomorrow, to see if there is any reply to my message. By the way for someone who's got English "just to get by" you're doing all right.' Vincent remarked with just a little sarcasm.

'Oh, I had to learn it, good for my business you know?'

'Aye, I wish my Spanish was as good, I knew it just to get by, but that was a long time ago, adios señor Manuel, nos vemos manana!' Vincent uttered in Spanish with a rather comical accent.

'Gracias señor Vincent, a manana,' the worker replied.

After coming out of the post office Vincent crossed the square diagonally, towards a building that had the "HOTEL" inscription on it. And he wasn't even surprised to see his companions seated on the terrace, already checking out some of the local beverages, although still early in the day. Vincent took a little detour before being noticed, not willing to say anything about the message he had just sent to Washington. Not yet, not before getting a reply so that he could have a better picture before talking to the lads.

'We found a place boss, not really the five stars, but hey we're in the middle of the jungle.' Wade Johnson said loudly as Vincent stepped onto the terrace.

'Yeah I can see that, not too early for drinks either, I've noticed.' Vincent commented before pulling up a chair from another table.

'It's nice and sweet boss, they say it's made with honey. They call it "Para", it's like cider, here try it for yourself.' Wade told him handing over his tall glass.

'Nah, maybe later, after we decide who is on watch at the boat for the night.'

'That's being arranged already, Clive offered to stay over.' Wade announced sounding amused. 'Clever move, he'll enjoy the rest of our vacation,' he added, more amused still.

'By the way boss, for how long will we hang around this place?' Running Wild asked.

'I don't know yet, perhaps three, four days at the most.' Vincent replied unconvincingly. He didn't want this conversation now, and it would be better to avert it at all costs for the moment, although he could feel the anxiety in the men. 'Did you meet the proprietor of this place, I might need a word, hopefully he is a westerner,' he said, pretending not to know anything about the place.

'Sure did, he is American, Mr Emmett Floyd, from Pennsylvania. He's made his fortune in the rubber industry we're told.' Wade answered, while sipping from his drink. 'For the record, we didn't pay anything yet, we said that you'll join us and then straiten that out,' he explained after taking another mouthful of the aromatic brew.

'I see, well he can help us in acquiring the supplies we need. Don't worry Wade, your drinks and all your other needs are covered by the federal government.' Vincent joked.

'Way to go boss, isn't that what we wanna hear boys?'

'Indeed,' the half Indian said while the doctor and the Englishman appeared to have no interest in such talk.

'I may have a chat with Mr Floyd gentlemen and then I will take some rest if you will excuse me.' Vincent told them as he got up and made his way to the lobby.

The evening arrived quite soon, as Vincent could tell after a few hours of good sleep. He crashed into his bed right after talking to the hotel owner, arranging for the supplies to be purchased, not even eating for the day. But now he felt hungry, but more desperately he needed a bath. The room was equipped with a tub behind fabric curtains opposite the bed. There were a water barrel, sponges and soap and even clean towels, all these seemed like they were from a different picture considering the location. He spent a good half an hour enjoying the lavishness of satisfying a rather basic need. Feeling clean and somewhat restored Vincent went down to the saloon only to find his mates, half drunk, even the Englishman and Dr Jacobson seemed steamed at this stage. More so Wade and Running who were accompanied by two pretty ladies, one a native of the place and the other more western looking, tall, blonde and very attractive. The men were in good spirits laughing and talking loudly, draped over the piano where a local boy was playing cheerful tunes. There were a few other customers, both locals and outsiders but even these appeared to be well settled in the place. Vincent made his way to his companions' table after scrutinising the saloon a little more.

'What can a man do to get some food around here,' he asked in a jovial tone to suit the general mood of the place.

'Just ask,' Wade said, 'see that beautiful lady over there, she'll feed you alright,' he gestured towards the bar.

A few minutes later Vincent was stuffing his mouth with a tasty chicken, garnished with aguaje fruit made from a local recipe, that he ordered from the beautiful bar tender.

'She is some ass eh? Mrs Floyd I mean..., good looking bird he's got there, that Emmett fellow.' Wade said quite spitefully.

'Here Wade, enjoy the lady on your lap and forget about it.' Vincent told him very abruptly. The fact was that Wade Johnson was known for losing his mind around women and if that wasn't much of a concern, Vincent recalled well the times in Cuba when Wade used to rape young women. This was unfortunate for him, but he was a good soldier and a hell of a shot nevertheless. However Vincent had to be on his toes and watch the man in order to avoid unwanted conflicts with the locals.

'What are you drinking there Mr Englewood?' Vincent asked more to divert attention from Mr Wade's earlier comment.

'Good old scotch, not very cheap in these parts but your federal government is paying, isn't that what you said,' the Englishman replied sarcastically.

'Aye, but I might invoice the Royal Society for your expensive taste though.' Vincent laughed.

As he spoke the hotel owner, Mr Emmett Floyd, handsome middle aged gentleman, tall and athletic, dark hair and deep green eyes, came around from Vincent's back, drawing patiently on his cigar.

'Is everything in order for you fine gentlemen,' he asked with a refined accent.

'All okay, dinner was delicious, my compliments to the chef!' Vincent replied pleasantly.

'I am happy to hear that, may I join you for a few minutes, I usually come around to talk to special guests,' the man with the educated accent said.

'Sure thing, please have a seat Mr Floyd. I guess you already made acquaintance with my companions so that's absolving me from making the introductions all around.' Vincent replied.

'Indeed,' Mr Floyd said as he sat right in between Vincent and the Englishman.

'Well you maybe forgot to introduce your lovely wife, quite a sensation around here I believe.' Wade Johnson spoke rather provocatively.

'Huh, Claire? Yes she is my wife but I don't get your meaning, what is "sensational" when it comes to her, Mr Johnson I believe,' the hotel owner replied sounding both confused and perhaps insulted.

'I am sure that Mr Johnson meant that as a compliment, am I right Wade?' Vincent answered in a rushed and severe tone.

'Off course, it was just a compliment, no offence Mr Floyd.' Wade said with determination, but for those who knew him better it was noticeable that the cowboy did not mean one word.

'None taken,' the young businessman said rather circumspectly.

'Gentlemen, I propose a toast for our two brothers lost in the wilderness, shall we raise our glasses for professor Mathews and for Mr Gonzales?' Running Wild suggested just so attention would be diverted and tension reduced.

'I've heard that you ran into some ordeals gentlemen, I present my condolences for your lost friends, to their memory,' the hotel owner said raising the wine glass he was holding in his right hand. They all followed suit and then were quiet for a few moments. Vincent was a little annoyed to learn that his companions had spoken about these things already, but also appreciated the half Indian's intention of trying to change the subject.

'Do you mind me asking, what is it you are looking for this far out in the wildest place on Earth?' Mr Floyd asked genuinely curious.

'We are conducting a scientific expedition, recording animal and plant species that may not yet be known to the world.' Mr Englewood lied. 'Like the giant snake we encountered, one that size was never recorded to my knowledge,' he said.

'Yaqumama, or else "mother of the water", that's how the locals call them. You must have been very unfortunate to run into one like that, but I am told if the rainy season is a little dry they come to nest near main bodies of water. They may not be on your records Mr Englewood, but they're well known in these parts. One that size can swallow a bull with ease, real monsters I heard.' Mr Floyd explained.

'We didn't run into it, we saw a boat abandoned on the river bank, the fucker must have made a few killings before getting our friend, we saw human carcases on the shore.' Vincent told the hotel owner.

'...again, sorry for your loss,' Mr Floyd said sympathetically,' how far do you intend to go on your journey gentlemen? I mean there is menace every step of the way and all this sacrifice in the name of science, well I find that hard to digest really. Is it gold you are after? Because I came across crews just like yourselves, searching for "El Dorado", the lost city of gold.'

'Funny you should say that, because you are the second person to mention this city of gold, we were asked this very question back in Manaus,' said Vincent. 'What's the damn story with it?'

'I think it's just a myth, but it's said that the "Gods" have built this city entirely out of gold, somewhere at the heart of the Peruvian rainforest.' Mr Floyd explained.

'The gods! They've built a city out of gold? Well shoot me dead!' Wade Johnson exclaimed.

'Let the man finish his story!' Running Wild cried. He then nodded towards Mr Floyd in a sign that he should go on.

'Yeah, the gods or sky people, you see the natives are very superstitious, full of these stories concerning gods coming from the skies, building things and so on. If you ask me they're only legends, pretty much nonsense, so is "El Dorado." If there was such a thing it should have been found by all these crews going in search of it, but they never return, what a waste of life,' the businessman told them. 'What direction are you taking gentlemen considering that we are at the big junction? If the city of gold is what you're after everyone went up the Maranon River, turned right so to say,' he added.

'Not at all, we're going left on the Ucayali, so we are not after the city of gold Mr Floyd,' Vincent said confidently.

'Well what do you know? Anyway I wish you best of luck finding your species, just keep safe and enjoy the rest of your stay. If you'll excuse me I have more customers to attend to.' Mr Floyd concluded sipping the last mouthful of wine from his glass.

Shortly after Mr Floyd's departure Vincent gave Wade a disgusted look but decided not to talk about it, not with a man who was drunk and well known for his violence. They spent another while in the company of beverages and spicy lady companions. The next day Vincent went to the post office first thing but he learnt that there was no reply to his message, which only increased his anxiety levels. He still hoped that two more days in the town was plenty of time to receive it and therefore went about his business, buying necessary supplies for their journey. He was pretty nervous about talking to the men regarding the continuation of their journey, whatever they would decide though, he still needed them to bring him to the drop off point at least. After the conversation they had with Mr Floyd the previous night, it had kind of awakened his interest in relation to their quest. The Englishman appeared even more excited speaking with eagerness about the local Peruvian folklore and the sort. However Vincent recalled him saying something like myths are called myths for a reason and by no means should they let themselves be led by such stories. On the other hand Running Wild had a different take on these matters and he told stories that he was taught in his early years by the Hopi elders. Their own culture was full of "sky people" and so on, that had contact with their ancestors. Vincent found these accounts bizarre indeed, but even the English professor admitted that there might be truth at the bottom of these stories. However for the moment Vincent left the philosophical matters for more practical ones. Another day passed and still there was no reply from the president, not even late in the evening as he checked just before closing time. Señor Manuel told Vincent that if any message got through overnight, he would leave it over in the hotel's reception after learning that Vincent planned to depart early the next morning. As it was the last night in town, they were all in the hotel leaving the boat to be guarded by a local man, introduced to them by Mr Floyd, who also vouched for him. It was the time for having a word with the men and for Vincent the pressure was mounting with every minute as it got closer to that moment, which inevitably arrived. He met them all in the saloon at around eight in the evening, and as per his instruction they all gathered in a corner at a table that was somewhat isolated from the rest.

'Good evening gentlemen,' Vincent said as he joined them. 'I think you all know why we're here or shall I remind you. A few days back, after Mr Gonzales died, God rest his soul, I've asked you to reconsider your thoughts about this expedition. You're all free men, so you can choose to risk your lives further, or you can turn around and head back to America, back to your families, in any case I need an answer now,' he added gravely.

'I thought we were contracted by the federal government, should we choose to go back would mean losing our reward, am I right?' Clive Orson said enquiringly.

'That appears to be the case indeed Clive.'

'A man who tumbles over hardship, ain't no man to me but just an innocent boy, for one thing, and then there ain't much for me to return to but the county jail. This is freedom boss, so I ain't going back unless I do what I came here for. I'll be better off a dead hero than a rotten prisoner. This mission is the only reason they let me out, besides I am no coward,' cried Wade.

'I gave thought to it boss, what can I say? If I go back now I don't get the rest of that money, my boy can't follow his dream, he is good with books you see? He wants to go learn the law up in Massachusetts. Now the government promised to pay that money no matter if I am dead or alive. Well that brings comfort to my heart knowing my boy can follow his dream, so I will stay by your side boss.' Clive Orson explained.

'Me too Colonel, I worked way too hard all my life, couldn't save a dime. If I am to die, my family will have comfort living in wealth at least. Let's go find these "gods" and who knows, we might get lucky and get back in one piece,' said Running Wilde.

'Alright gentlemen, you show character, each for his own reasons, how about you Professor, Doctor Jacobson, what do ye say?' Vincent asked the two men who hadn't spoken so far.

'Well, speaking for myself I have higher purpose than just the money, this is a unique quest and truly I think it should be concluded,' the professor said.

'Um, I can't say how much use I was so far, you know I couldn't help Mr Mathews back there, but I believe you still need a doctor around, so I will go all the way.' the doctor spoke casually but with some regret in his voice.

'Now, I can't guarantee that we'll all make it, even if I wish it, but this jungle has proved to be unforgiving of the slightest wrong step that we take, but honestly gentlemen we can't allow ourselves to be laid-back and ignorant anymore…, no more errors, we need to watch every step of the way without rushing, no faltering in front of any threat, because there is only danger out there, waiting every step of the way. We'll be on foot for many days after the drop off point and that's going to be a real bitch, you hear me?' Vincent urged.

'Loud and clear boss, let's do it,' Clive replied confidently.

'We'll leave before dawn gentlemen so take it easy on the drink. We'll meet on the dock at four o'clock in the morning, sharp.' Vincent told them as he got up. They all nodded, not very convincingly, perhaps not liking the advice as regards drinking. After all, this was their last night in a place that resembled civilisation, to some extent.

Four o'clock in the morning arrived awfully fast, for Vincent at least, who was now aboard among the others, with the exception of Wade Johnson whom nobody had seen since after midnight. Vincent was a little concerned about this, but he decided to give it a few minutes before sending one of the men to look for him. He went over and checked the supplies again, for which they had spent rather a fortune as nothing was cheap in these parts, and inquired with Clive Orson about the repairs being done to the boat. He was also bothered by the fact that no message had come back from the White House but he also knew that being President of the US was rather a demanding job. However after a short while he became disconcerted and considered sending Running Wilde in search for the "missing" man. As he stepped back on deck he heard distant shouts and something like a firearm being discharged. There were multiple voices, more gun shots which became clearer and clearer with every second. Vincent instantly became very nervous, he felt that whatever was happening was not good and likely something had gone terribly wrong for Wade Johnson. It was only seconds before the picture became clear as they could now distinguish Wade's voice.

'Start the boat now, hurry goddam it,' he was shouting from the distance, sounding out of breath and very distressed. He was running fast towards the docks, obviously being chased, as farther behind him numerous voices shouting swear words in Spanish could be heard.

'Start the damn motor Clive! Get the shotguns boys Mr Wade is in trouble!' Vincent ordered without hesitation. Instantly the boat's engine roared into life and began reversing as Wade Johnson landed on deck, his whole bodyweight hitting the floor so hard that the boat shook in that split second.

'Fire at them, fucking hell they nearly got me,' he said panting heavily.

Without delay two gunshots penetrated the air as Running Wild and Vincent both discharged their rifles in the direction from which the multiple voices were coming. Then it became silent except for the "Mary Rose" barking its engine as it pushed forward upriver. It only took a few seconds before more gunshots split the skies, only they came from the opposite direction, as the bullets hit the boat's hull.

'Fuck, push it faster they're too close!' Vincent shouted while firing his weapon towards a few human silhouettes that now took shape on the docks. More swear words, more gunshots, were being exchanged for a short while before the boat sailed out of range for the men on the docks, leaving their desperate screams far back in the distance. Nobody on deck was hurt it seemed, but Mr Wade's shirt was covered in blood.

'What the fuck happened Mr Wade, are you hit, you're soaked in blood,' Vincent shouted at the man who was still lying on the deck.

'...it went wrong man, it all went fucking wrong,' cried Wade, still catching his breath.

'What went wrong Wade, what did you do, are you hurt?' Vincent shouted once more, leaning over Wade's body, bringing the gas lamp near his face. The expression on the cowboy's face was that of panic and confusion but also Vincent could sense there was something else, a viciousness, as if Wade Johnson was acting out his emotions. Without a doubt whatever had happened was dramatic. However knowing Wade it was more than likely that he had brought this tragedy upon himself.

'It's not my blood, boss, something went so fucked up boss..., and I had to kill him.' Wade said, controlling his breathing. 'I am sorry but it was either me or him,' he added.

'Who? Who did you kill Wade and why?' Vincent asked in a severe tone, kind of guessing what was coming.

'I went for her boss, that Claire women twisted my head like no other and then he came in, the fucking husband came in with a pistol. Luckily I had mine just within reach boss and I managed to pull the trigger first. I am sorry it wasn't supposed to turn out that way.'

'What about her, did you kill Claire along with her husband,' cried Vincent in pure desperation.

'I did what I was supposed to do boss,' the cowboy replied calmly.

This news made Vincent's body tremble, transcending his entire being, in a pure state of shock, he threw himself backwards, allowing his body to lay down on the cold deck while feeling waves of heat and sweat pass over him.

'Here Wade, you committed crimes that not even God could or would forgive.' Vincent whispered almost sobbing, after a short break.

'Well boss, ain't I blessed that I am on my way to meet Him? I mean this is what our quest is all about, isn't it? Finding God! And when we do, He'll be the judge of all that I've done, all the crimes and all the sins I've committed, whatever they were.' Wade replied rather hastily.

'Yeah Wade we're here to find Him and if we do, judge you He will.' Vincent spoke with determination.

Chapter Nine

'Madame et Monsieur, Ladies and Gentlemen! May I have your attention please.' this gent was shouting out loud upon the stage, desperately trying to cover the frenetic cheering from the abundant crowd, 'We have the lovely and the unique, the beautiful and the spectacular Claaaaire Floyd,' he ended as the crowd erupted in applause so loud that Vincent instinctively covered his ears with his palms.

'But would she survive the might of the beast, the one and only elite shooter Waaaaade Johnson,' the speaker shouted, which brought the cheering to absolutely insane levels. 'Would she make it through on the spinning wheel, please Ladies and Gentlemen give her a round of applause,' he said as two young men pulled down the black curtain at the right hand side of the stage, only to reveal a wooden wheel on which a half-naked beautiful woman was stuck, in an unnatural pose, her legs and arms spread as far apart as her body would allow, forming an "X" shape. What really frightened Vincent was the expression on the woman's face, which was that of sheer horror. At the opposite end of the stage this cowboy came out running and jumping all over the place, performing forward and backward flips with great ease, as if defying gravity. He was dressed in immaculate white shirt and trousers with black boots and a white hat with a large borders. The shirt and trousers were embroidered with all sorts of patterns, elegantly, arranged throughout. But his face was in complete contrast with his white outfit; it was dark and expressed rather a terrifying evilness that would make one's skin crawl instantly. He was also holding two silver pistols with long barrels, nicely crafted as if custom made for this evil show. The two young boys began to spin the wheel harder and faster each time and then suddenly they jumped to the side while the cowboy shot his two pistols simultaneously. Nobody could tell whether the poor woman on the wheel was being hit, because the cowboy wasn't even aiming at anything in particular, he just kept on shooting towards the wheel. The crowd were completely out of control by now shouting all sorts of slogans.

'We want blood, we want blood,' they were all shouting in one voice.

'No, no Wade, don't kill her for God's sake!' Vincent found himself whispering. But as he spoke something even more dramatic took place. The top of the large colourful tent split open while above a shining disk like object began spitting thunderbolts right into the crowd, each bolt hitting one person at the time, transforming them into ashes instantly. People were running chaotically in all directions, screaming and shouting in pure terror, attempting to escape the menace. Up on stage the wheel was still spinning at amazing speed but there was no sign of the shooter anymore. Instead the speaker was left in the middle of it laughing hysterically as if this entire horrific episode was a comedy for him.

'I knew He'd come for you, you are all sinners, there is no escape now,' he was saying out loud in a sarcastic, wicked tone.

Vincent was frozen, not able to move a muscle, not even his face. He stood there, turning his head around in all directions, to see the whole town being bombarded after the tent entirely collapsed in flames around him. He began screaming for help, digging his finger nails deep into the wooden chair in front of him. He did it so hard that when he looked at his fingers, they were bleeding and painful. He was sweating, for he felt the heat of pure hell was now taking his breath away. He closed his eyes, squeezing the eye lids shut so tight that tears instantly ran out. Shockingly when he opened his eyes he found a completely different picture in front of him, no more flames of hell, no more thunderbolts, no more people turning into ashes, but just a bare grey timber wall with his finger nails digging into the end of his bed.

'Fuck me!' he barely spoke, with difficulty trying to regulate his heavy breathing. He got up, contemplating his bleeding fingers. It felt hot and stuffy inside the cabin making his head hurt as if he was experiencing a massive hang over. Slowly he stepped outside onto the deck where the heavy tropical rain was hitting the floor so hard that it produced a deafening noise.

'Are you still mad at me boss?' Wade's voice came as if out of nowhere. Vincent turned in the direction from which the cowboy's voice came. The man was wrapped up in a bulky cloth rain coat on which the heavy rain drops were making a strange noise.

'Forget it Wade! Just fucking leave it alone!' Vincent replied in pure disgust.

He wanted to go back inside his cabin and even made a few steps back, but then suddenly he decided otherwise. He walked towards Wade instead, feeling anger rising from deep within.

'Did you know the Floyds had two small children Wade? Did know that?' he asked staring right into Wade's eyes. The cowboy looked down towards the floor in a sign either that he felt sorry or perhaps he was hiding something, but Vincent suddenly realised it might have been a much worse scenario.

'Oh no, you didn't Wade, tell me you didn't you mother fucking murderer!' Vincent shouted, at the same time he sent a heavy punch straight into the cowboy's face making him tumble backwards.

'It got out of hand, the little rats went crazy, screaming, and I panicked that's all.' the cowboy shouted back, spitting blood from his injured gums, 'Fuck you broke my tooth,' he swore.

'I should rip out your heart, you son of a bitch, but you ain't got one now, have you?' Vincent swore back at him. He wanted to launch another punch but was stopped in his tracks as Running Wild caught him from behind, encircling him around his shoulders, automatically restraining his every move. Vincent winced as he felt the pressure on his shoulders and on the back of his neck. Likewise Clive Orson jumped on Wade rendering him immobile in an instant.

'Enough boss, we all feel that way, but there is no point, besides we may still need him,' the half Indian shouted in Vincent's left ear.

'You can let go now Running, I am done here, I am done with you Wade, you twisted fuck!' Vincent cursed as he broke loose from Running's grip.

There was nothing of pleasantry exchanged after this, there was nothing but this savage rain that was beating against the "Mary Rose" deck, becoming very heavy at times. The mood amongst the men aboard was significantly low; perhaps the lowest since they had got onto this boat and you couldn't say that they weren't challenged before. This felt different because, whatever challenge they faced before pausing in the town of Iquitos, might well have been classified as accidents, but Mr Wade Johnson murdered a whole family in cold blood just because he lusted after Claire Floyd and that could not be classified as anything but pure evil. Vincent felt the heavy weight of this event because he employed Wade although he knew him better. He knew of the many times that Wade had violated and killed innocent women back in Cuba. He simply had no excuse to have signed him up after all that, and now Vincent could hardly bear the guilt he felt. He certainly played a crucial role in bringing the Floyd family to their untimely death. As for Wade Johnson, perhaps he could put a bullet through his head alright, but that wouldn't change anything. So what was the alternative, just let God deal with this? It would be odd considering the purpose of their mission here in South America was to find God. According to the account given by the Scandinavian explorer they were going to find

God in some physical form. This idea of finding God as a physical entity was hard to contemplate as it didn't make any sense, not to Vincent, because right now he could only accept the theological concept, whereby one could only meet the maker in the afterlife, when judgement would be made upon them. Vincent needed to believe in this option, just to feel comfortable that Wade Johnson would be punished for his horrendous crimes. For many days Vincent just beat himself up, not being able to shake these thoughts off, not finding any peace whatsoever. The only thing that made him feel somewhat better was his conviction that by bringing this expedition to an end, perhaps justice would be done with regards to Mr Johnson, whether it was in this life or in the afterlife. They were getting closer to the drop off point now and the crew was feeling apprehensive, more and more with each mile that the "Mary Rose" was making upriver. The Ucayali River was getting narrower and the flow was getting faster, it felt, slowing down their progress and making the boat's motor work harder. On the upside, the supplies aboard seemed to be lasting a while and mainly they were good quality items, which Mr Floyd had guaranteed back in Iquitos. How ironic that, just after making a good sale he never had a chance to benefit from his profits. Nevertheless Vincent had paid a fortune for all these products and, even worse, it was hard for him to see Wade getting intoxicated on good quality whiskey after just taking the life of the very man who provided them with the spirit. There were now only a few hours left to go on the boat and, according to the maps, the drop off point should be in sight just a little over five miles away. It was clearly inscribed on the map and it appeared to be a long narrow beach on the right bank of the Ucayali. It was early morning, July 3rd 1912, and the skies were finally free from the heavy clouds that had dominated for many days. The river surface was foggy, steaming up under the warm rays of the sun that were now falling gently onto it. As far as one could see the ground fog penetrated deep into the forest on both shores. It was unusually quiet as if the whole place was asleep, except for the boat's engine strokes which could be classified as noise, and at times flocks of birds could be heard taking off in the distance as if disturbed by the engine. Running Wild was the only person on deck, deeply preoccupied with putting some sort of device together. He had a long rope on which he was hanging two empty tin cans every so many inches, tin cans that he had carefully collected all along this journey. Up inside the cabin Clive Orson was steering the boat, as he took over from Wade. He was concentrating on the navigation but he could not but notice the passion and the skill with which the half native was working on his device.

'What's that for Running,' he shouted down to his colleague.

'Will keep us safe at night, you see when we make camp I will surround it with this rope, set it at different levels above the ground so no intruder will come in unannounced, so to speak.' Running Wild explained.

'Aye, that's clever,' the new skipper praised him. 'I think we close to that damn beach there Running, so if you wouldn't mind to shout 'em all up..., lazy scum,' he added with sarcasm.

A few hours later all the men were on deck, already packed up for the journey on foot. It felt terrifying to lose the sense of security that the boat had offered, not always, but most of the time, however from this moment forward they were going to be totally exposed to harm. This was in addition to their dampened spirits, as the Iquitos incident was still affecting them, less perhaps for Wade Johnson, the perpetrator. But right now at that moment, that particular event mattered less and it would be wise maybe to put it to rest. The green immensity ahead of them, surely one full of nasty surprises, was more of relevance, and even Wade Johnson was more caught up with it than his recent macabre crime. The gear needed for the journey was supposed to be light, so it would not impede progress to their destination, however after seeing it all on the river bank it was quite a long way from that concept. There were food supplies, water, weapons and ammunition, tents and other tools considered to be vital, including a rather bulky snapshot "Brownie" camera, which seen some use so far. It may have seemed excessive because they were also two men down from their initial complement. Vincent walked around the stuff laid out before him, in no particular order, trying to assess whether some gear could be left behind on the boat.

'Food,' he said, 'I think that's too much to carry around, besides we can hunt some beasts if need be, then rifles and pistols, one of each for every one of us and yeah two tents must do. One man needs to stay put, guarding the camp at all times and that will include you doc and the professor.'

'What! They don't know anything about guarding a camp boss, with all due respect.' Running Wild jumped as if being burnt by Vincent's suggestion.

'They will learn, everything they need to know, and you folks are gonna teach 'em just that, am I clear?' Vincent replied sharply. He then got down on his knees beside the rifles, picked two of the lighter ones and threw them one after the other towards the two men of books.

'Clive, you show the two gentlemen how you use these, do a little target practice before we set off into the forest, we're down one goddam good shooter, so they better learn quick, cause we'll need them,' he suggested.

'As you say boss,' the tall cowboy replied, sounding a little confused and then he spat out, over his right shoulder. 'You heard the man gentlemen, so pick 'em up and stop looking down on them as if you never seen one before.'

The doctor and the professor performed the same motion all at once, picking up the rifles in an almost comical way, not being sure what end they should hold the weapons from.

'They will shoot us dead boss, look at them, they must be dummies, never held a gun before!' Wade Johnson laughed out loud.

'That will do no harm Wade, if you're the first to go down.' Vincent told him, vengeful still.

'Boss is still mad at me, damn it, boss got soft, like he never seen death before, it was a damn accident...!' Wade commented in passing.

Vincent did not pay any attention to this, instead engaging in tidying up the gear along with Running, while Clive Orson was giving instructions on gun use, to the best of his abilities, to the two educated gents who were literally lost before such a lecture. It was only Wade who found nothing useful to do and therefore kept on laughing and making unsuitable jokes. It all came to an end and after a good hour's work the six men were now making their way slowly into the depths of the jungle. The pathways varied from very narrow to non-existent and add to that the level of concentration and anxiety, there was minimal progress being made. They were heading in a slightly south-eastern direction as indicated on the maps provided by the deceased Scandinavian explorer. Running Wild was leading the formation cutting through the thick vegetation, using either a tomahawk or a machete in some cases both. His dexterity was to be envied, swinging his arms around him, describing imaginary circles round his body and hacking at bushes, lianas and all sorts of plants with fascinating ease. For many hours this was the only noise to be heard, that of plants being cut, along with the sporadic flutter of flocks of birds or monkey squeals in the distance. The six men kept quiet, did not exchange one word, trying to remain unnoticed as far as possible. Also idle chatter in this environment was just an unnecessary waste of energy. The jungle was dark and humid making them dehydrate with every step of the way, while fatigue was setting into their bones awfully fast.

'We'll pause for a little while Mr Monroe, please? We're getting tired.' Dr Jacobson suggested in a whisper.

'Aye, only for a little while, night will soon be upon us, and we need a safe place for camping.' Vincent said louder so that the whole group would hear him. They slowly gathered in a circle and sat down on the moist ground after making sure that no living thing was in sight. They shared water and they caught their breath whilst trying to clear the sweat that was blurring their vision, with their dirty sleeves.

'How long is it until we reach our destination boss?' Running Wild enquired.

'According to Mr Rasmussen, God rest his soul, about five to six days to the edge of the forest and perhaps another two to get up onto that plateau, but at this pace maybe more.' Vincent explained.

'Mysterious are the ways of the Lord, aren't they folks?' Wade asked, his voice sarcastic.

'Do me a favour Wade, shut the fuck up, you have no Lord.' Vincent thundered his words.

'Well boss! You're not this saint you try so hard to portray either. You killed people too and many...' Wade replied, spitefully this time.

'Never without reason Wade, I've never killed anyone without reason...'

'Oh yeah, and what's that? Who gave you the right to decide on whatever you call reason, How do you choose boss, who dies and who lives? How do you know, for the ones you killed, you had reason, because I got to tell ya, it makes no difference between those I killed and those you killed, reason my ass!'

'You probably right there Wade, it makes no difference, we both killed and that's that, can we settle this now?' Vincent responded in a calm manner.

'Hell yeah! You the one so mean boss, ain't me!'

Shortly after this bitter exchange of opinions, the group was on the move again, finding it more and more difficult to progress. The terrain was rising slightly, the jungle was getting even thicker in places and darkness was fast approaching. After just a couple of hours Running Wild spotted a place that appeared to be good enough for making camp. It was an area of about two to three hundred square yards, almost vegetation free apart from some tall grass, in the centre there was a solitary tree, not very tall, but with a large crown. There was still a short while left before complete nightfall, so being in the open, the area was bright enough to permit the setting of a proper camp. Without delay the tents were erected near to the solitary tree, after a thorough inspection had been carried out by Running. He also surrounded the place with his tin can rope "alarm system" and they all decided to have one man on guard at all times, by taking turns for two and a half hours each. They decided to use the food that they had brought with them from the "Mary Rose", one because it would save them the time and energy needed to try and hunt, but also because it would make sense to get rid of some of the weight. Thankfully the night passed without incident and gave them a chance to get good rest for the day ahead, when they would hopefully be able to go farther and much quicker now that they were used to the kind of obstacles the forest had to throw their way. As daylight slowly began replacing the night, the men were in high spirits, more optimistic and certainly motivated to carry on. They had coffee and dried bread for breakfast, not necessarily the most appealing menu but any source of energy was beneficial. The only inconvenience at this time was their water supply, which got significantly low, so most of them agreed that finding a clean stream was a priority, however there was none indicated on the map, not near enough anyway, perhaps a bit over a day or so distant from their current position. Thankfully, Running explained, streams are not the only safe water source in this environment as most plants would contain considerable amounts of water in their stalks, as well as rain water lying on top of large leaves which could also provide a safe source of hydration. With that piece of security in mind they went on for the next day and then the day after with no significant occurrences taking place, except for many cuts and bruises which they all acquired while squeezing through the thick vegetation. Injuries which doctor Jacobson was well equipped to treat and thus not allow infections to spread. In spite of all the positives they were still not making much progress and on day four they had just covered about thirty to forty miles, less than half of the distance to their destination and certainly a lot less than what they have hoped for. It was just about midday when they arrived at another opening, on high ground from which point they could see the plateau in the far distance. But between their location and it, there was still a great distance to cover, of hills and valleys of thick forest. This was heart breaking for Vincent, anticipating that maybe another four to five days at least were needed until they reached those plateaus.

'Fuck, I had enough of this green hell, I hoped we were much nearer to the edge of it,' he commented just about to cross over into the open area. In that instant he felt a firm hand on his shoulder, squeezing it to the point of agony almost.

'Don't take another step!' Running Wild whispered from behind. 'I feel that we are being watched,' he added.

'By whom or by what?' the group leader dared to inquire, as at that moment a strange noise penetrated the air as if a small object travelled at very high speed. In an instant doctor Jacobson fell down on the floor, hitting the ground like a sack of potatoes, making a deafening sound on impact. A small arrow was sticking out of his neck, with blood spurting all over that area. The man instinctively brought both his hands up to try and take out the foreign object that was causing him excruciating pain, but his body rapidly went into convulsions and before anyone could act, doctor Jacobson became inert. They all went down on their knees trying to take cover using the vegetation. It quickly became clear that they were "face to face" with some locals and unfortunately they were exposed, unlike their attackers. There was not a soul in sight, but judging by the way the doctor was hit by the arrow, the "shooter" must have been placed directly across the open plain, using some sort of blowing device, which also meant that they were close enough, perhaps just on the edge of the tree line upfront. They needed something fast, a solution to save their lives, but the surprise was so overwhelming that it made every one of them remain frozen still.

'Mr Wade, you good at aiming your rifle from your hip, am I right?' Vincent whispered out his words.

'Yes boss! That I am!' the cowboy replied from nearby.

'See that colourful bird at the top of the small tree; shoot the damn thing so they can see we have fire power too.' Vincent suggested. It was the only thing that came to mind, intimidation tactics which in theory at least should work, considering that their weaponry was far superior to the aggressors'. However shooting from the hip was no easy task, the procedure itself was dangerous and only a few shooters were capable of executing it in the right manner, without getting injured. Wade Johnson was one of them, one who had practiced and perfected this manoeuvre for many years. It takes a very steady hand and a clever way of distributing your strength so that the body can take the force of the rifle's recoil. Wade placed the rifle on its side, running across some of the length of his forearm, and the stock with its butt firmly stuck against his upper arm, right above the elbow joint. He then lifted his forearm to about forty-five degrees, placing his elbow tight against his hip and putting a significant amount of his power into his right arm and abdominal muscles. He slowly tilted his head so that the corner of his right eye would be aligned with the end of the barrel, while the barrel itself would be directly aligned with the target. The whole procedure took several seconds, but only in this way could one have the confidence to carry out this complicated shooting technique. In a split moment the gun thundered through the air, echoing for several seconds afterwards in the far distance, while the bullet hit its target right in the chest making the colourful bird literally explode. There was silence once the echoing noise of the rifle disappeared among the thick vegetation of the surroundings. Vincent slowly came to a standing position, not knowing exactly what he was to achieve but more hoping that if he showed no further aggression, they may come to some peaceful resolution with the still unseen "enemy". He lifted his rifle up in the air and stood still for many moments, but nothing really happened.

'Come on up all of you, do what I do, lift your weapons up in the air,' he whispered back towards his mates.

'Are you insane boss, they will kill us all,' cried Wade, who was nearest to him. A moment later Running stepped right beside Vincent, imitating the surrender gesture just like his boss.

'He is right, it is our only way to stay alive, so get up boys,' Running said loudly.

It was almost spectacular to watch, as the tree line across the small plain became alive with countless human shapes coming into open view. There were about twenty individuals of small stature with dark red skin, warrior like painted faces and bodies with hardly any clothing at all, except for the genital area where a piece of animal skin was covering their private parts. They were all holding some sort of a weapon from spears to long pipes to very primitive bows or animal bone knives. They slowly came towards the middle of the plain whispering words that no one could make out. One of the small men, perhaps the leader, came close to Vincent, holding a spear in a menacing way, indicating that Vincent should drop his rifle. With no second thoughts Vincent pushed his right hand forward and let the weapon follow the gravity. All his colleagues followed suit almost instantly.

'What now, Running? What do we do next?' Vincent asked sounding very uncertain, perhaps blaming himself for his earlier judgement to surrender.

'They're likely to take us to their village and also they're likely to kill us there, so we only delayed our deaths gentlemen', the half Indian spoke gravely.

'No fucking way one of these midgets can take me out!' Clive Orson cursed through his teeth.

'Whatever you do, don't do it now Clive or boss is dead,' Running replied.

Moments later, the men were all stripped of their belts, pistols and knives as well as all the other equipment they were carrying. Also they all had their hands tied together behind their backs, with some sort of strips, perhaps made out of tree bark, whatever it was, it was a hard material cutting deep into the skin and flesh around the wrists. The three mile or so trip to the locals' village was completely agonising as they were pushed and shoved from behind with each step of the way. The village itself was very primitive, consisting of several huts set out in a circle of about two to three hundred yards radius. In the centre of this perimeter there were all sorts of objects, laying around in total disorder, such as cooking pots, fire wood, water basins, most of these looking ancient as well. There were mostly women, children and elderly scattered over the yard; they all stopped whatever it was they were doing, just to stare at the new arrivals, literally speechless. But what appealed to Vincent in particular were the many pillars erected all around the place, every one of them displaying colourful symbols hard to describe. Vincent looked straight out to make eye contact with Running Wild perhaps the only person who could interpret the drawings. Running nodded, not only because he understood Vincent's visual message, but also because he had already considered a plan in order to avoid torture and even extermination.

'Boss do you still have the map in your pocket,' he shouted over his shoulder.

'I think I do but I can't reach it,' the leader cried.

At that moment Running Wild began shouting, mostly sounds with no meaning, not to the English speakers anyway. A few of the local men came close to Running, in an intimidating manner once more and one of them even hit him with his fist, right in his abdomen. The half Indian crawled in pain but kept on shouting. It was consternation that could be read on most locals' faces. The women and the elderly, even the young ones began screaming in return, turning the entire surroundings into pure chaos. It all stopped as this old man, perhaps in his 80's, came out of a hut, farther apart from all others. He was a small man, more so than the rest of them, bald but with facial white hair, wearing an animal cloth-like skirt, just about covering his genitals. However he had several ornaments all over his chest and neck probably each of them symbolising something in their primitive culture. He walked slowly and did not stop until he reached close enough to

Running's position. He then visually measured the half Indian, from top to toe and back a few times and eventually said something that had no meaning to any of the foreigners, unless perhaps for Running Wild.

'Do you know what they say; you understand 'em, Running?' Clive whispered, as if not to attract attention.

'Just about, but I can't speak back, I just shouted words that I heard them saying,' Running replied in a whisper as well, 'just don't speak before he's done,' he said.

The old man stepped back a couple of yards and nodded as if giving Running a chance to speak. All his companions held their breath at this point, realising that the next few seconds would decide whether they lived or died. Running remained quiet and just made eye contact with the elderly man, then he turned to the left and showed his hands being tight at his back, came back to his initial position and then turned his head towards Vincent nodding hard. The old man looked over to Vincent and yelped while gesticulating with both his hands towards Vincent. A few local men ran to Vincent, two of them holding spears to his face while another began untying his injured wrists.

'Take out the map and put it at my feet, then slowly reach for your sack and take out the notebook from the Scandinavian explorer, do it!' Running said calmly.

In slow motion Vincent reached for his left shirt pocket and took out the folded piece of paper and then slowly moved towards Running, where he unfolded the map putting it on the ground. Every single soul was now encircling the foreigners, making them feel overwhelmed. Vincent looked over to the old man and pointed his finger out towards his sack, left idly on the side. The man nodded back to him indicating that he could attend to his bag. After a quick but cautious search Vincent returned with the note book and placed that beside the map, opening it at the pages full of symbols. The elder came close and looked carefully at both items. Running Wild allowed him some time and thereafter he began repeating just one word over and over again. The elder tilted his head backwards looking right up to the sky and then he lifted his arms towards it. Running kept on nodding and pronouncing that same word which sounded like "taio'ua" or close enough to it. At the same time the old man began shouting a word as well, but sounded far from whatever Running was saying. Next thing they knew Running had his hands free and once he could move he bent over the note book showing the elder the symbols and then pointing out the colourful pillars. It was in this instant that Vincent realised some of the symbols and pictures drawn by Mr Rasmussen were quite similar to some of those inscribed on the pillars. Minutes later, they were all freed and let lose to walk around the village; however men with spears were following them closely.

'I never doubted you one bit brother.' Wade Johnson addressed the half Indian.

'I am not your brother Wade, never have been and never will be.' he answered with disgust.

'Oh, what are you now, their brother?' Wade replied provocatively.

'Cut it of Wade,' Vincent said sharply, 'what now, what are we doing now Running?'

'We are free to go I think, but may be wise to spend the night here, there's no safer place and we need rest.'

'Is it safe though?' Wade interrupted.

'It is for us but I don't know about 'em with you hanging around.' Vincent responded with an air of bitterness. He then walked away not listening to any of the nonsense Wade Johnson had given in reply.

Chapter Ten

What is a man's life worth, especially that of man who has no regard for life anyway? The life of a man who took countless lives of others with no hesitation, without remorse and even more had no consciousness in doing so. Probably worth nothing, on the contrary, if such man is put to rest now and forever, it just means that no one else need fall victim, for this man is only a predator, a merciless killer for whom only God holds the answer, but only then, when he passes into the afterlife. It was horrible what had happened to Wade Johnson, some explained, but Vincent was hardly making any sense of what he was told. He couldn't make any sense because he had hardly any recollection of any events since they halted at the local tribe's village, for their minds became clouded with visions, terrible visions and horrific nightmares that could only be explained by the strange plants, the smoke of which they began inhaling, ever since that day, the day they were captured by these natives. And none of them could even remember how long ago that was, perhaps days or even weeks, or God knows, but surely it had been some time. Today it was different because Vincent woke up in the midst of chaos as people, both locals and foreigners were all screaming and shouting, and he could not understand why. However he did understand that it wasn't yet another vision because he felt soaked and when looking down on his shirt and at his hands he could see a lot of blood, still fresh and moreover it did not appear to be his. What had happened then?

He was told that Wade Johnson was found dead in the woods; just a stroll away from the village and Vincent had blood on his hands, a lot of it. So the maths were simple, Wade dead and Vincent soaked in blood, still though, how about the mechanics of such an occurrence, if there wasn't even a faint memory of his having committed this crime? With his head still confused and incapable of putting up any resistance, he found himself being dragged across the yard, by some of the locals and his companions. He was then dropped on the ground in front of a woman, who was naked and had blood on her face and body and she was held with her arms and legs spread against two pillars, one on each side, all her limbs being tied to the pillars. Vincent only became more confused in his attempt to make any sense of what he was experiencing. He desperately tried to put things together but his mind was very blurred and except for some dim images which incorporated a period of time that he could not approximate, there was nothing but this present moment with the anarchy unfolding all around him. He looked up at the woman with blood on her face only to be terrified by her expression, which was incredibly cold, her eyes blank with her eyeballs withdrawn into her head, all in all a picture so horrifying that Vincent felt ill.

'It appears that you killed Wade, early this morning when he was over fucking this girl, but you don't recall it do you?' Running's familiar and friendly voice sounded at the back of Vincent's head.

'What..., I don't even know how I got here, what's happening to me?'

'It's okay; they will help recover your lost mind, just don't fight it whatever it is they do,' the half Hopi man whispered.

Whatever happened after this, it is hard to explain in words. But for sure it cleared Vincent's mind and helped him travel back in time, as if his mind was spooled back to the moment they had arrived in the locals' village. Strange images flashed before his eyes while the locals were feeding him some weird tasting leaves and shouting out even weirder incantations that he could not make out. He could see himself getting intoxicated on smoking mixtures of strange plants and drinking exotic local beverages, giving him and his companions' extreme pleasure but followed by great sickness. Even odder was the way he could see these pictures now, as if he was an outsider who witnessed all these events from the side. It all culminated with the event that happened only hours ago which left Wade Johnson lifeless at the edge of the village. And yes it was him, Vincent Monroe, who followed the cowboy who was dragging this young girl into the nearby bushes, dropping her on the ground and before he knew it, he had stabbed him in the back, over and over again, until the man had no breath left in him, and until Vincent found his release in sorrow, until all his regrets for bringing Wade along on this expedition were dispelled.

'I killed him, I've stabbed him in the back Running, that's what I've done,' he said, almost breathless after regaining consciousness.

'We knew that, we just wanted you to remember and it's okay now that you did. It was perhaps God's will to punish Wade in this way for all his crimes.' Running told him.

'As if we need God to punish anybody when we can do such a great job ourselves!' Vincent uttered with regret. 'And now what, am I being judged by the locals, am I next for punishment,' he then asked, almost panicking.

'No! I think they just want us gone, but there is a problem, the professor is ill and can't move,' the half Indian informed him.

'What's happening to her, why is she being tied between the pillars?' Vincent enquired as he got back on his feet.

'She is part of a ritual, I guess, they see you as part of a maleficent deity and she has been touched by it now, so she will be sacrificed.' Running explained in whisper.

'No, that can't be, tell 'em Running, she ain't done nothing wrong!' Vincent shouted out.

'Sorry boss, we can't hinder their ways, their beliefs, don't you see?'

'If they see us evil how do they regard you Running?'

'I am the only reason for us being still alive boss, now we need to get going, no more questions!' Running Wild urged in a demanding tone.

'How about Clive, where is he?'

'Getting a stretcher ready so we can carry the professor, we figured if they head back to the boat, there is some medicine and other goods left there, you and I can finish the journey together,' the half Hopi man explained.

'They can't, it is not possible for one man to go through the thick forest dragging a stretcher with an invalid..,' cried Vincent.

'Two of the Pa'Wa warriors will accompany them, just to make sure they get back safe.'

'Can you trust them?' Vincent asked sharply while getting his back pack ready.

'We're still alive, that's something, and besides I don't have a better plan. To carry the professor along with us, chances are we'll never make it.' Running said shaking his head.

'You may be right Running, can't see any alternatives either, I just think we got away rather easily, after all that.'

'They were happy for us to leave behind all of Wade's belongings, including his weaponry, they never seen anything like it. I showed 'em how to use it. It will make hunting much easier at least for a while. We promised more ammunition from the boat; Clive is to give 'em that once they reach the "Mary Rose".

'Sounds fair, let's get going then!' Vincent said with a sense of enthusiasm in his voice. He had to admit that if Running's plan was not ideal, it was a life saver at least and for now they had no better alternative. After giving a few instructions to Clive and saying his goodbyes to the ill professor, Vincent was about to step out of the compound, making his way through the huts, when the elder's voice called him from behind. As he turned Vincent saw the man standing in front of the pillars where the young girl was still tied. He said nothing, just gave Vincent a look that cut deep through his soul. He then pointed his finger out to Vincent and slowly raised his arm to the sky with his finger still stretched out. This was the last image Vincent ever had of the elder. The next few hours that followed were intense as Vincent and Running Wild kept on going deep into the forest, perhaps less cautious than ever before, but they covered a fair bit of ground. Vincent was amazed how the Pa'Wa people had indicated a pathway that would put him and the half Hopi back on track to their destination.

'How those primitives could read maps,' he asked as they halted.

'I am not sure but there is more and perhaps they are not as primitive as we think. Okay they don't have roads and factories but they have deep knowledge.'

'Meaning...?'

'For instance they know the sky, the stars and they know it well. If you looked over to those pillars, the way they were spread around, it matches a certain portion of the sky which Europeans call Orion, in my mother's tongue it is called "Hotom'gam". The fact is that the Hopi people moved across the American south-west for centuries building their settlements in direct arrangement with those stars, they think it's the place of birth, and according to Mr Englewood almost every culture in the world shares this idea. So does the Pa 'Wa. And if this isn't fucked up enough, they recognised our destination on the map, it is a forbidden place for them but you see, they actually never travel that far out anyway. It's just that their ancestors told them about this location, "the place of the gods" where no man should enter.'

'How long was I gone into darkness, what the fuck had happened to me?' Vincent asked almost as if ignoring Running's talk about the Pa 'Wa's culture.

'Nine days, boss, you were intoxicated with "vision herbs" and drink, brewed from the same plant. I was too, but only half as much. Then the professor fell ill and I looked after him. I hope he'll be okay if he lasts the journey back to the boat.

'I thought you didn't like him...!'

'Aye, he's not a bad man, maybe a bit ahead of himself, arrogant sort, but not bad.'

'Well Running, you are a blessing boy..., so we'll head to the "forbidden place where no man should enter".' Vincent concluded.

For three days in a row the couple kept on walking through what seemed to be an endless forest, camping wherever they found it safe, facing scourging heat and heavy rain, but nothing would stop them. Eventually the pair came close to the plateaus which were now within sight but still with some ground to cover. They agreed to keep going until dark and that perhaps would nearly take them out from the depths of the rainforest, with all its splendour and dangers, all it's fascinating colourful wild life and deadly predators and all that the largest forest on Earth had to throw at any human who is stupidly courageous enough to venture this far. It was just about time for a sigh of relief as the high plateaus were getting closer with every mile covered, but this forest wasn't done just yet, not now and not without claiming the life of another. With all of the enthusiasm that they felt, the pair lost concentration as Running Wild went farther ahead and did not notice the gap that opened up at his feet. An eighty foot drop hid a small river at its bottom, the very place where the body of the half Hopi, half Irish man was laying lifeless, smashed against the slimy rocks as Vincent was staring hopelessly from above. There was no premonition of this sudden death of his companion, usually a man of great awareness, but it happened, quickly and with no warning, but the gap and its river where clearly indicated on the map drawn by the Scandinavian explorer, as Vincent was to learn a little while later. But right at that moment there was nothing he could have done to prevent his friend's abrupt death. Instead he cursed the gods, the forest, the president and everything else he could think of, for all he knew he could have been the only one still alive from the entire crew. But he also knew that this was the point of no return and only going ahead might hold the chance for him to survive this expedition. He went down the steep cliff and burned the body of his friend, right there on the river bank, instead of burring him. He also retrieved some useful items that the Running was carrying, amongst which Vincent found some interesting notes that the half Hopi took while talking with Professor Englewood. These concerned theories that the two men shared in relation to what they might find once they reach their destination, but also some pictograms they perhaps copied from the Pa'Wa's richly symbolic culture. Vincent carefully placed these in his bag for he found them of importance. As well he thought of the snapshot camera for it would be the one thing that can hold evidence beyond doubt, only to find it smashed to pieces as Running had it in his sack. Later on that day he made camp and decided to write in his diary, something he had omitted to do for quite a long time.

It's sometime in August 1912, I lost count of days and from all I know I may be the only one left alive from the entire group, perhaps Professor Englewood and Clive Orson are too, but they are probably back on "Mary Rose", although the professor was very ill. Only hours ago I lost a dear friend, Running Wild, he fell down into a hollow, maybe a good eighty foot drop. His body was distorted, unrecognisable almost, as it pounded against tree branches and rocks on its' way down. It happened suddenly and I could not have prevented this from happening as I was several yards behind him. God bless his soul, for he was a good man, but this ill place has no mercy, neither for the good nor for the cruel. Speaking of which, we stopped at a local tribe's village, the Pa'Wa people they call themselves and they took us into captivity at first, but then treated us as guests. Nonetheless they got us intoxicated with some weird stuff, smoking and liqueur of some kind that would cloud one's mind and give one horrible visions. It appears that while intoxicated I have taken Wade's life for he murdered people, young and adult, in Iquitos and he was indenting to claim more victims from the Pa'Wa people. I killed him in cold blood and I don't count myself as a murderer and I have no remorse for his death. I will let my crime be judged upon my return, if lucky enough to make it back.

Leaving all these ill fated occurrences aside, I am now at the edge of the forest, having the rocky plateaus in sight, which I hope to reach by this time tomorrow. I feel lost, discouraged and not one bit looking forward for what is next, but can't go back now, if anything I have a better chance to survive by going ahead on my journey. A journey that I feel should have never happened for we were on a quest to search for "God" and now I think that He has abandoned us. Therefore how can He be found? So far it's been nothing but hell, so what are the chances that God lives among such an ill setting? My friend Running and the professor seem to have different understandings on such matters and as noted by Running the possibility of finding "God" in this place is quite strong. He wrote this during the time that we were guests in the Pa'Wa people's village and added some pictograms that they likely saw around the village. These people, the Pa'Wa have great beliefs in the so called "sky people" and they claim to have been part of a rather great civilisation that became lost hundreds of years ago, called Tiwanaku, who used to worship a god by its name of Viracocha, who appeared to have lived around here.

But "here" is a vague notion because the vastness of this place is unspeakable. Running further noted that the comparison between the Pa'Wa's beliefs and his mothers' ancestors is rather striking. He further writes that according to Professor Englewood these beliefs are very alike with what he found in India while looking through old Sanskrit texts (as he calls them), where all sorts of sky deities were being worshiped, some good, some bad, some bringing prosperity to people, some others bringing wrath, destroying everything in their path through means that are hard for the tongue to speak of. Is there any truth, or are all these just stories that survived the test of time, being told from generation to generation? I don't know, it's just what I read in the notes I found amongst Running's belongings, God rest his soul (I think I wrote that already). I am the one with perhaps the least knowledge on this matter and ironically I am the only one left to conclude this expedition whatever it is that I am going to find, if anything at all. So I will get on my way, first thing at dawn, for there is still some road to cover and surely it will be far from danger free. I need therefore to use caution like never before for I am all alone in this endless wilderness and thus it is all up to me to keep safe, if there is any chance of that. I surely hope that these few notes are going to be seen one day, by those who sent me here, which means I have to stay alive and hand 'em over myself, or else nobody will ever know what happened to us, unless Clive and the professor return to the States. But even if they do, they can only tell the story until we separated and if I am to die (which is likely), the rest of story will die with me, same as it died with all those who already perished during this expedition, each would have had a fragment of the story that can never be told from their own perspective. What a shame!

Enough with the words for now, the night is young yet and I need to make sure that I'll stay safe for the remainder of it, but I shall scribble more whenever I get another chance, it is good for my sanity.

Signed: Colonel Monroe

Chapter Eleven

The scenery was changing with every step that Vincent was taking. The "endless" jungle did have an end and now from the luxuriant vegetation that one could barely penetrate, the place became rather plain with hardly any tree in sight. Instead there were bush- like plants, all that Vincent could see ahead, on an arid and rocky land that appeared to become endless too. High hills and valleys surrounded the entire area, it was perhaps much easier to travel on foot than in the dense forest, but he had to face a whole new set of challenges. For a change the place was very dry with hardly any living creatures around, for there was no water source nearby, and staying hydrated and fed would become quite a task. Furthermore his own provisions were coming to an end as he decided to carry as little weight as possible, just a couple of cans of beans and rice left, less than a litre of water and nothing else, apart from a pack of cigarettes. However the biggest challenge was his deteriorating state of mind in this desolate place. The scenery in itself was not necessarily unfamiliar, but this was not the States, this was not Texas neither Arizona, but the Peruvian pampas. This was a place hardly known to any westerner, few dared to adventure in these parts of the world and consequently there was very little information that Vincent could refer to before his departure. From up above on the arid plateau he took a quick glance back to the jungle he had left behind and by God, it was far more impressive looking from here than he'd ever imagined. The dilemma he was facing was that he could not really decide which was worse, the green hell left behind or the dry yellowy-brown rough terrain ahead. It didn't matter in the end because more important now was to try covering some distance and hopefully finding a water source. Better again would be if any sort of creature that was edible might cross his path and come right to the mouth of the barrel of his rifle, for he would not hesitate to shoot it dead. With that in mind he made good progress on the rocky landscape but soon fatigue took its' toll once again. He looked at the map while catching his breath and noted that the "forbidden place of the gods" was indicated right ahead, about forty miles, but in this environment forty miles to cover on foot was a fair bit. He hoped to have reached his destination in a couple of days, tops, but perhaps his hopes were rather illusory and not anchored in reality. He then thought that ten miles a day would be a more realistic target however if he didn't come across water very soon surely death would be the result. What was more the map drawn by the Scandinavian explorer showed no creeks or lakes anywhere near, but only the marshes near the destination where these pyramid-like structures were supposed to be erected. In spite of feeling tiredness setting in deep into his bones, Vincent decided to continue at his pace until nightfall. He could only manage just a couple of miles more and then suddenly his body fell to the ground, completely outside Vincent's control. He struggled to get back up but to no avail as he could only get back on his knees. *'Kneel and pray'*, he thought, because there wasn't anything else he'd be capable of. He struggled to pull out a blanket from his backpack as he was well aware that places arid such as this were very deceiving, while hot during the day, night time could easily turn into a freezing hell. He drank the remainder of his water; he really needed it although he had not yet located any water source, and the next thing he knew he was laid down on the floor being wrapped up in his blanket. Rapidly he lost touch with reality and the depths of his mind took him straight into "dreamland" right back to his early years as a young boy, right in the front of "The Olde Sweet Shoppe" on the main street in Brownsville, Texas. He was there with his father following the same old routine, buying sweets after the Sunday mass and by God they were good and they were plenty for him to chew upon. All sorts of sweets, from candies to jellies, from crunchy to soft with all the flavoured juices dripping down from the corners of his mouth, on his chin and even farther down on his immaculate white shirt that his mother had put on him, fresh that morning. She didn't mind Vincent dirtying the damn thing because she was happy to see him enjoying the long awaited treats. His father would feel that he had fulfilled his duties for the day, and so he'd head of to the "Saloon", there to drink whiskey and smoke cigars with fellow army men while talking through important matters that arose at the fort. Vincent was left therefore to his own devices in the care of his mother, playing battlefield games and that kind of thing, running up and down the hill nearby the fort, "shooting" his wooden rifle, "killing" many "enemies". But that always made him thirsty, very much so and now was no different as he felt his tongue was very dry, while sweat was pouring down all over his face and body making his shirt stick to his skin, which made him feel great discomfort. It was that terrible feeling, a mixture of severe thirst, sweat and dirt that brought Vincent back to the present moment, a present that for a few seconds he could not determine but it almost didn't matter. Priority now was to find water and nothing else, but there appeared to be another problem. As he woke up Vincent became disoriented, he could not recognise any of the landscape mainly because he had never taken any notice of it before collapsing. The position of the sun did not help him either because it was right above his head, so the only thing he could do was to search for his own footprints and establish which direction he had come from. A few yards away he could clearly see the marks and so he began walking in the opposite direction while trying to fight the cruel sensation wanting to urinate. He thought that if he did, his body would lose the last drop of fluid left and then he'd no longer be able to regain any. But then it struck him, what if he did, and kept the urine to drink so it would ensure hydration for another while? The thought of it made his stomach instantly sick, but what were the choices? Little to none it seemed. Without thinking twice he opened his water container, pulled a piece of cloth he had in his pocket over the container's mouth and tried to keep his penis steady directing most of the urine into the water vessel. The damn thing stank sending a sharp sensation to his nostrils. How about the taste of it? He decided to close up the container and perhaps wait until it cooled down somewhat, before drinking it. With some of his strength restored after the long deep sleep, Vincent went on walking towards the southwest, not at the same pace that he had the previous day, but he felt any progress at all would be of benefit. The landscape was becoming more desolate with every mile that Vincent managed to cover, making him realise that the chances of finding fresh water were now very slim. This was one thing for which he was now missing the jungle, for it at least was abundant in water sources, between streams and creeks as well as many of the plants that could hold good amounts of water in their stalks, there was never a shortage it seemed. But here it was so much different, and the only means of getting fluids back into his body was his own urine that he was holding in his water container. After severe efforts Vincent could no longer withstand the harsh thirst sensation. With his stomach crawling inside him, he opened up the container, covered his nose with one hand, and then brought the bottle up to his mouth with the other, drinking every drop of it. It was sickening, the salty taste, the warmth of it and everything else made his whole being feel repulsed, shivering violently trying to fight off the nastiness of the whole thing. It seemed that for the initial few seconds the feeling of thirst was in fact becoming more acute, and combined with the sickening feeling, made Vincent think that he was after having made a big mistake. Then shortly after the struggle, his body began to feel a little more comfortable with it and it no longer fought to reject the urine. Vincent felt his energy levels coming back up, shook his head as if to get rid of the thought that he had just drunk his own urine, and then quickly wiped his mouth with a clean piece of cloth before throwing his rucksack on his back once more, becoming even more determined to cover some distance. Ironically after just a couple of hours dark clouds quickly gathered above and within minutes it begun lashing down, fresh clean water, so much of it that he wished he had a few more bottles with him to fill up. Although it was cold rain it did not stop Vincent taking off most of his clothes and dance all around the place, for he found that he felt clean again and it made the drinking urine incident a thing of the past. On the other hand he was only lucky to get the rather short lived rainfall, as it does not happen very often in these parts of the pampas. After the entire struggle though, the bit of rain was just a blessing. With his morale high Vincent walked a good distance until nightfall, crossing hills, valleys and plains, the whole landscape the same rather disheartening scenery. When he woke up the day after, on the hill top where he made camp, he discovered a very uncommon view, far down in the distance. In fact Vincent could not believe his eyes as what he was seeing appeared to be four pyramid like structures, erected on a plain surrounded by high hills about four to five miles away. He quickly took out his monocular and the map, and after having a good look at the structures he realised that it was his destination. In between him and the site there were a number of smaller hills that he should have no problem gaining in just a few hours, but what then? What was he supposed to do when he reached the site? Because he was not an "expert" in any of this and the few people that had some knowledge of this matter were nowhere near, moreover one of the professors if not both, were now dead. At first glance the place looked to be deserted, but he was still too far away to make that assumption. Therefore he decided to try and get as close as possible, perhaps onto one of the smaller hilltops ahead, and spend a good bit of time just observing the place before making a decision on how he'd proceed further on. With that in mind Vincent packed up his rucksack, attended to his physiological needs and slowly started to walk downhill. He was very nervous and had this bizarre idea that he might have been noticed, the question remaining, by whom? Or by what? As a trained soldier in a situation where he needed cover, he'd know well how to use nature to his advantage, but on this bare terrain camouflage was not much of an option. Instinctively he kept a low body posture, walking sideways at times, which made him tired awfully quickly. Out of breath, Vincent decided to halt for a little while, took off his back pack and dropped to his knees. As he tilted his head back in order to stretch, the most spectacular event started to unfold right in front of his eyes. Just as if the sky above had cracked open, this large oval shaped object slowly began descending. The object was surrounded by a bright purple light and in that instant several thunder-like noises covered the entire area. Multiple bright orbs came out from various parts of the large object and after dropping in slow motion they all began travelling at high speed, in no particular flight pattern, but rather chaotically all around the place. After the initial shock Vincent slowly came back to his senses, only to feel this massive panicky sensation taking over. Rapidly he grabbed his backpack and his rifle and looked about in all directions trying to find a hiding spot. The only reasonable location he could see was straight ahead, uphill, where a rock formation was sticking out near the top. With no further consideration he ran right up to it, dropped his bag, kicking it deep under the rocks and then threw himself on top of the rucksack, taking cover in between the formation. He quickly weighed his options, only to figure that there was no way to go from here unless he wanted to give away his presence in the area. The good thing was that none of orbs flew near his current location, while the downside was, he could no longer see what was happening on the other side of the hill. For a few minutes he could only manage to stay put, clinging on to his rifle, squeezing it tight with both hands while trying to fight off the overwhelming panic. He could not think straight, he could not imagine what he would do next, as if his mind and body were being crushed by this unique and horrifying feeling. Slowly though, Vincent's brain appeared to come back to reality and slowly too his body felt less stiff, allowing Vincent to consider his future options. The first thing he did was to take a quick look behind his hiding place only to see that many of the flying orbs were descending right on top of the pyramid structures, one by one disappearing inside the structures. A few were still flying at high altitude in a rotating pattern while the big oval shaped object appeared to be hovering high above. The bright aura-like light surrounding the object was changing colours at regular intervals, every tens of seconds or so, from bright purple to yellow, red and green. Moreover there was lots of steam coming out of it, from several parts, almost beginning to form a cloud all around it. It was hard for Vincent to describe what he was seeing but instinctively he took out his leather covered note book and a couple of pencils and quickly started to draw the phenomenon. It was very similar to the pictures drawn by the Scandinavian explorer long ago. Even so, Vincent glanced over his own sketches and realised that no matter how much he tried to draw the phenomenon accurately, it was far from the real thing, which literally defied any sort of description, as if the whole thing was unreal, some sort of hallucination that one could experience. It was very bizarre, Vincent thought, and certainly any education he had had on religious descriptions of "God", was nothing like this. His own mother, a deeply religious woman, spent lots of time teaching him about such matters, often reading out parts of the Holy Book, on late evenings in the winter, right by the fire place. Vincent still recalled some of these "stories", like the one about Moses and the "bush on fire" or when the Israelites were guided out of Egypt by a "pillar of fire" and other stories of that sort, He could only imagine these things back then, just the way they were described and he had thought for long enough that these descriptions were rather absurd. Mother always told him that the "word of God" is not to be doubted and he had better believe what the "Holy Scripture" was saying. But there it was, he was here now in the middle of Peru all on his own, sent in a "quest to find God" while witnessing events outside any imagination. Had he found it? Because with all his lack of knowledge on such matters, what he was seeing in front of his eyes was very advanced machinery, flying machinery; the kind that humans don't even dare to dream of. He once saw an aeroplane, on display in Brownsville, not long ago. It was there for people to see the real progress of humanity and as Vincent recalled it, it was made of wood and cloth-like material with metal wires all over the place. At the time he thought that no machinery could ever become more advanced than that wooden aeroplane, but here, whatever he was seeing right now, call it whatever you like, there was no doubt that it was technical in nature and millions of years more advanced than a wood and cloth aeroplane that all of his fellow humans talked of as being this majestic wonder of human technological progress. But then quickly Vincent's mind went to a time, for just a moment, when he was a young boy and his father handed him a book, nicely covered, that had many pictures inside, pictures of technological marvels, out of this world. It was an adventure book written by a French mad man who imagined a projectile capable of reaching the moon and coming back to Earth. What an insane story he thought at the time, but captivating nonetheless. There was some degree of similarity between the phenomenon Vincent was witnessing and the French novelist's story, at least in principal, Vincent believed. Deep in his own thoughts, Vincent suddenly felt an unusual heat that surrounded the entire vicinity and was accompanied by a bizarre noise that he sensed more like a vibration. Vincent could not decide whether all this was just in his head or maybe something was happening just outside his hiding place. Something told him that he should stay put and not move a muscle, but this insane desire to find out what it was that generated the heat and the vibrating noise, was getting the better of him. Hesitant still Vincent held the rifle close to his chest so that it gave him a sense of security. However fate took an unexpected turn as this flying object came right around the rock formation and began hovering at low altitude right in Vincent's face. He had no more cover which was terrifying, but now that he was exposed he had to act. He jumped up onto his feet, still holding the rifle close to his chest but he remained frozen, still not knowing what else to do, first because he had been taken by surprise and second because he was facing something so unfamiliar that neither training nor experience could have prepared him for such an encounter. The flying object was a cone shape, not very large and had some wings attached to its fuselage which were changing geometry every few seconds. It was made of metallic materials, bright silver with a glass case at the top coming right down to the nose of the apparatus. Inside it there was a being, human-like, looking like its face was covered by some sort of a mask, out of which a thick hose and several other pipe-like objects were appearing, somehow connected to the apparatus itself. The noise from the flying object could now be heard at low frequency, making Vincent's ears hurt. It wasn't loud but it was a sort of deafening hiss, spreading out in all directions, generating air currents that Vincent could barely withstand. He instinctively raised his rifle, pointing it right at the being inside the object but quickly realised that his mighty gun may not be worth a dime in front of this super advanced machinery. On second thoughts, the being inside was probably unprotected since he was only surrounded by the glass case. Without hesitation Vincent pulled the trigger generating a thunder-like noise after aiming right at the being's head. To his great surprise the creature inside did not even attempt to take cover as the bullet ricocheted off the glass case, which appeared to have released some sort of bright blue electric spark on impact. The panic and fear inside Vincent's entire being reached alarming levels in that instant. He quickly dropped the rifle on the ground and then followed suit by dropping down on his knees with his hands raised above his head in a sign of surrender. During his tormented life, Vincent had been in several situations in which he felt paralysed by fear, back in the war and even on this expedition, but now it was different because he could not see any way out. Running away from it was not an option while fighting it was even less of an alternative. The sensation of being paralysed intensified, not only generated by fear but also because some sort of weird hawser came out of the nose of the flying object and rapidly wrapped around Vincent's body, catching him tight round the waist. Shortly after this a second wire caught his legs, right around the ankles, they both sent powerful electric shocks into his body, making Vincent feel very stiff. His mind though remained sharp, perhaps he was very confused but he could clearly see himself being lifted up in the air with his arms flung downwards. The sensation was one of pure agony to say the least. The object flew forward for just a few yards, halted and then another cord-like item with some sort of anchor at its end stretched down picking up Vincent's backpack and rifle. He struggled to escape but the grip was so strong that he could barely shake his body, moreover with each attempt to do so, powerful shocks were sent all through his body and the agonising pain was now making him physically sick. The overwhelming sickening feeling made Vincent pass out, for a short time, as the object turned around swiftly and flew over towards the pyramid-like structures. As he came back to his senses Vincent saw an opening at the top of one of the structures through which he began to descend slowly, as the cables were lowering him down until he touched the floor. The descent was smooth but that made him sick again and as he came up on his knees, after being freed from the restraints, he could no longer help it and began to throw up, mostly some watery fluid with a bad taste. It took him a few seconds to fully comprehend that he wasn't alone, and the place was brightly illuminated. His vision was somehow foggy but he could distinguish several silhouettes all around, tall beings moving in the most peculiar way, almost as if they were defying gravity. A black globe like object, which was floating up in the air came within a few feet of Vincent and instantly sent out rays of bright blue light that were going up and down his body for quite some time. He could not make out anything that was happening and as his vision began to clear, Vincent saw large panoramic screens, right up on the walls of the structure, which were displaying all sorts of pictures and symbols, some of the pictures appearing to be those of the hills and valleys nearby. What seemed even more peculiar, most of the screens were showing these pictures in motion as if something or someone were screening the entire area right at that moment. Vincent could not make a lot of sense, either of the live pictures or of the symbols that were displaying repeatedly, every so often, arranged in some system of writing that for him was unreadable. Even stranger still, there were pictures of the moon, it seemed, on some of the screens, taken from near it. Throughout the structure there were several desk like objects, where tall beings in peculiar clothing were standing, appearing to watch very carefully these thin rays of coloured lights which were taking the shape of very large globes, some certainly representing the Earth, as Vincent could make out the shape of the Americas and the African continent. Some others were not showing the continents but rather appeared like groups of smaller spheres arranged in some sort of pattern that reminded Vincent of the constellation that his now dead friend Running was talking about. But none of this artificial imagery of Earth and other places mattered for Vincent; he was in lot of distress right now, surrounded by strange biological intelligent creatures that certainly would have had something in mind for him. Kill him maybe, that would certainly be the easy way out and the one Vincent surely wished for. The only thing Vincent was sure about was that the next few minutes would decide whether he would pass into the after-life or continue to exist in this life, which felt beyond surreal. He was finding it hard to tell if he was imagining all this, like in a waking dream, as if his mind was playing games. Then there was the pain he was experiencing all over his body caused by the uncomfortable flight he had taken while suspended by cables, and also the sickening sensation he still felt deep down his stomach. These feelings were undoubtedly real and so it appeared that all these beings, every one of them at over seven feet tall, were as real as they could be. One of them came very close to Vincent, looked down on him and extended his arm towards Vincent in a sign that he should get up to a standing position. It appeared to be a male, bald with an unusual elongated skull and a most peculiar skin colouration, a mixture between purple and pale grey with deep dark blue eyes, no facial hair, dressed in a peculiar grey suit and silvery boots. As his right hand was still stretched out towards Vincent, he noticed that the being was counting four unusually long fingers instead of five. Three more creatures came close also, encircling Vincent. The most frightening aspect of the entire appearance of these beings was that they had no expression on their faces, but seemed blank, as if they had nothing to communicate, no emotions to show, nothing at all but impassive dead cold.

'*Homi Nuuus*', this high pitched voice pierced the air, coming from behind the being standing right before Vincent, '*ak-ta Te Hominy*' *(look at you human)*, it continued, echoing all around but also sounding closer. As the tall being standing before Vincent stepped back, Vincent saw this other creature coming straight at him, moving in a bizarre way, almost floating rather than walking, but also undulating its body on both sides as if unseen forces were pulling it from left to right and back again. It appeared to be of female gender, wearing some sort of a long dress, metallic blue in colour. The being was as tall as her male companions, presenting most of the same characteristics, as in the skin colour and elongated skull but covered in what seemed to be hair, nicely pulled up from the forehead, going round the elongated part of the skull, also the four fingers and also common to all, there were some strange symbols tattooed on their faces in different patterns from one another. Although very alien, Vincent found this entire appearance, was attractive in the sense that her body shape and facial features were harmoniously organized. Other than that she looked and behaved rather eccentrically, which Vincent found almost repulsive. She came near to him so that she put her outsized palm on Vincent's forehead and looked him right in the eyes while saying something hard to make out, often tilting her head backwards looking up at the large opening at the top of the structure. More bizarre were the sort of giggly sounds that she was making, overall this entire behaviour was hard for a human being to comprehend and even to tolerate. Many times Vincent tried hard to escape her touch which appeared to make the creature laugh even louder. Although this carry on did not necessarily feel threatening, it was strange and made Vincent feel extremely apprehensive. The other beings spent their time just watching the entire scene in a rather passive way until the male creature standing in front of Vincent shouted something hard to comprehend, however some anger could be sensed in his voice.

'Na Skod 'Hos a Man 'Us (*no harm to the human*), the female replied angrily as well. She then bent over Vincent and put both her hands on his face squeezing softly on his cheeks. 'Homin-us num-ki Te A-ka Na Kame 'Lom, Te Nah Hominy Mal-Coo-Tah, E-te Dheu 'Nos!' (*The human will now join us in the heavens, the human will not return to the kingdom of man, only death for him there*) the female creature said straight into Vincent's face as if he was supposed to understand and talk back. Instead Vincent stayed unmoved as if frozen looking at the being in pure disbelief.

'Who are you people? What do you want from me', he barely managed to say after a few seconds. The whole group of beings gathered around him and giggled in the most bizarre way. This was a hell of freak show, Vincent thought, but right at that moment this was just adding to a hell of a load of other problems. It was the uncertainty of what the next moment would bring that was getting to him, as many times before he had found himself in dangerous situations, but he'd always have a clue as to what to expect, even if that meant death. Well here and now not even death could be the anticipated option. Feeling totally powerless and on the verge of psychological meltdown he just had to abandon himself to fortune, leaving everything up to these alien beings which were now seeming to ridicule him, stripping him of any sort of courage or pride that he might have had before he was dropped inside the structure. These beings were nowhere near what he was taught about the whole concept of divinity, nowhere near the kindness and compassion that his own mother had preached when talking about God, and pretty much nothing else resembled any sort of perception that he had in relation to these subjects. These were freakish beings, not from this Earth, which happened to be far more technologically advanced, but who felt evil and cold, had nothing of holiness and for whatever reason came from a far distant world in machines hard to imagine. They were rather mean and harmful and, Vincent sensed, they wanted to bring harm to humanity. So the Scandinavian explorer must have been wrong, for he found not "gods" but quite the opposite. Were they devils then? Something was telling Vincent that they were neither. With a little more adrenaline surging through his system he came up to a standing position which only made him realise the true stature of these beings. By God they were tall, built to perfection judging by how well-proportioned they were. No human would ever stand a chance in a head to head fight with any of these creatures. He looked around measuring every one of them and saw no difference as if one was the perfect image of another and not only because they were the exact same build, but there was nothing to suggest any age distinction either. How that could be possible was something that Vincent had no answer to, as in fact he had no answer for anything.

'Are you going to kill me now or what,' he said, somewhat louder, his voice echoing around the building.

'Ne Dheu 'Nos Homin-us, Ne Dheu 'Nos Te'. (*No killing, we won't kill you*) the female creature replied in that instant, as if she perfectly understood what he said, 'Te Na Rog-ha High-Yea, Trans A Te Na Mal-Coo-Tah,' (*We'll take you away from the kingdom of man*) she then said, her voice kind, Vincent sensed. The creature came in very close proximity to him, invading his personal space once more in so far as her body was touching his for a good portion of it.

'Te Na Mal-Coo-Tah A-ta Nor-rah, A Te A-Ta Me Un Sun 'Us Sus Na Kame 'Lom. Un Sun 'Us, E-La Jota 'Sei A Man 'Us Qor 'Os E-La Dei 'Wos Qor 'Os. Na Homi-ni Mal-Coo-Tah E-La Che Sna, E-La Nah Dwe 'Nos! Te Homi-ni Muds 'Tos Ty-Ah-Woo-Tha, Muds 'Tos Woi 'Da Te Woi 'Na,' (*You won't return to your kingdom for now, but you will give me a son up above in the heavens, a son that will resemble both, your kind and ours. Humans will bring harm to their kingdom, that's not good! Humans must be given teachings, they must know retribution*) the creature ended, after whispering much of all that she had said.

It did not make any sense whatsoever, whatever she said was spoken in a language that Vincent had never heard, it sounded very bizarre and did not resemble anything familiar, apart from the one word that the creature used in different variations a few times, the word "hominy", which was repeated a number of times in different forms. It sounded near enough like "human" and in Vincent's limited linguistic knowledge it resembled something of Latin origin. He knew that because he spent much of his life amongst people that were native Spanish speakers. But whatever she said, although not understood, sounded quite important and he had the feeling that it mainly referred to his near future. Moreover she gave him an unfamiliar sensation, a mix of fear and excitement, a very strange vibration as Vincent perceived it. It was almost sexual in nature and there was a strong scent of lavender, which got imprinted deep into Vincent's brain. Overall this weird closeness made Vincent feel extremely uncomfortable, many times over from whatever he had felt up to this point. He became even more confused now as the beings stepped aside, while this strange silvery globe-like object floated in the air nearby, positioning itself within a few yards of Vincent at eye level. It looked as if the strange object had eyes as several bright red dots appeared to be staring at him. In an instant the object let out countless rays of red light which kept going up and down Vincent's body, similar with the way he was "searched" just earlier by the black orb. After just a few seconds the floating globe made a weird noise, a kind of screech that sounded metallic in nature and reverberated all around the area, and then it opened up on one side, to Vincent's left, letting out an extension that looked like an arm. In that instant it began sending out multiple bolts of lightning which hit Vincent straight in the chest, as powerful strings of energy surrounded his entire body, making him fall down onto the stony floor. From that moment it all went dark, no fear, no excitement and no adrenaline rush, not even the lavender scent in the air, but pure numbness and blackness.

Chapter Twelve

When he returned to his senses, the first thing that struck Vincent's brain, was that he was feeling cold and wet. Moments afterwards he found that he was naked, laying in what seemed to be a puddle on top of a cold metallic surface. As he opened his eyes fully he saw that he was contained in a rather large enclosure with obscure illumination except for an opening up above from which some light was coming through. Also there were a number of cables and chains hanging right from the top, sporadically banging into one another generating screeching metallic noises. Drops of cold water were coming down continuously, but not in large amounts, it was more like light winter rain. Some sort of rumbling noises could also be heard in the far distance, outside the enclosure or even further away from whatever this was that contained the enclosure. But ultimately and even scarier was the realisation that whatever this "thing" was that Vincent was inside, it was certainly in motion, very smooth indeed, but this "thing" was moving nonetheless. *'Oh no!'* he thought, while a severe shiver went through his entire being as he came to understand that the "thing" could not be anything else but one of the large "sky vessels" that he had seen some time ago, hovering at high altitude over the pyramid structures. Without even realising it, Vincent was now standing with his feet covered by the cold water on the metallic floor underneath. It was pretty dark within. He could see just a few yards in all directions as his eyes were still at the point of adjusting to the poor illumination. But he could distinguish that the area had some dark walls not very far away, arranged in a curve as if the entire room was a large sphere. On a closer inspection and with his vision adjusting more to the obscurity of the place, he noticed that the walls were filled with dark holes in the shape of irregular hexagons, large enough to comfortably fit a well-built man, however Vincent could not tell how deep these pits were. It made the whole appearance of the place look fairly similar to a beehive, but it also added a frightening dimension to it. One could only imagine what purpose these holes would fulfil and Vincent could only think the worst, although he had no indication of what that might be. He stood there for countless moments more, not being able to work out what he should do next. He was naked and cold, so cold that his body began to shake in slow tremors at first, getting worse with every second. He instinctively hugged his own body but it didn't make much of a difference. His mind remained sharp still, but something told Vincent that perhaps this was not going to last either. He wondered to himself where he might be taken, as he concluded beyond doubt that he was on what he now called the "sky boat". He quickly recalled the story written by the French mad man, about the few people that went to the moon and then returned to Earth. Was that his destination? The moon, or perhaps farther away? His father once told him that there were other celestial bodies apart from the moon, well away from Earth and that people named them after different gods that they believed in, back in ancient times, such as Mars and Jupiter, but other names too, which he could not recall right now. Were these other far away moons the possible locations that the strange beings were coming from? Was it there that they were now going? Hard to tell and it wouldn't make much of a difference since he had been taken away from his home planet. The more relevant question would certainly be, why? Why was he taken, because surely they did not want to kill him, they could have done that a million times over by now, but was that still to come? Certainly being left in this dark place, with no clothing and almost frozen, he wasn't going to stay alive for much longer. Why was he dropped here naked? As he asked himself that very question the strong scent of lavender hit his nostrils with a vengeance. This happened, not because he was imminently close to any of the creatures, but the strange sensation came from deep within. He instinctively looked over most of his body and started to touch many parts of it with his hands, only to notice severe bruising and swelling that he could not remember acquiring any time in the past. These were fresh and had no explanation as to how they got there. Suddenly flashes of memory struck his eyes, painful images these were, making him fall down to his knees and cry out loud. He could see himself laying on his back while that female creature, smelling like lavender, was taking pleasure in him. *'God,'* Vincent cried *'please tell me that it is just my mind playing tricks on me! That is it; I am just confused, my mind lingering on things that never happened,'* he lied to himself almost whispering his thoughts out. He felt sick once more and tried to vomit but there was nothing to come out. He was no longer feeling cold, on the contrary he was now sweating. Still in denial at least in some part, Vincent could only conclude that whatever it was they did to him was dramatic. Slowly a slight headache began to settle in, mostly coming from the back of his head, much stronger behind his left ear. He brought his right hand over and started to touch the area with the tip of his fingers, only to sense a small bump that felt bizarre. When pressing it, he could feel that this hard pointed area was somewhat sticking out, just under his skin. As he pressed on it harder something even more peculiar happened, he started to visualise sequences of numbers, zero and one only, randomly flashing in front of his eyes, line after line, so many of them that he could have filled out countless pages of it. *'Damn it,'* he cursed not knowing what to make of all this. He felt though that somehow those beings had inserted something inside his head, some sort of object that was causing him much discomfort and stress, making him see random numbers. What sort of sick things have they done to him, for what reason and what else would follow this evilness that he was being subjected to? These were questions that he could not answer, not even close to anything near the truth anyway, simply because he hadn't had the slightest idea about any of it. Also he was now trapped on a "sky boat" going God only knows where, or how far. Suddenly he remembered another story that his mother told him, a frightening story from the "Holy Book" where this prophet called Jonah was swallowed by a giant fish and where he spent three days and three nights right inside the belly of the fish. Vincent could only imagine back then how painful it would have been for any man to be swallowed by a fish, no matter how big a fish it was, but there again was it a fish at all? Because for some reason he found that story to be relevant to his own circumstances, as he felt truly swallowed himself, trapped inside the belly of this "metallic beast", being controlled by some maleficent creatures that do all sorts of bad things to men. And then the words of Mr Thorn sounded as clear as day in his head: *'make no mistake Colonel, but that jungle out there, you and your companions will be swallowed by it with no warning,'* whatever relevance those words had, if not by the jungle, swallowed he was, nevertheless. All these thoughts and all these questions, with no answers, could only generate a sense of hopelessness, one that Vincent had not experienced in a long time, perhaps from the time he abandoned his own fate and left everything to be taken over by the bottle. That was the time when he felt his pride just ebbing away, the time when he so quickly fell from dignity to depravity, only for this mission to have got him back to being a man, standing proud once again only to fall rapidly, right back into the abyss of pure humiliation, not by choice, but by the will of the "evil gods". Time and time again Vincent questioned the mechanics of life, how does it really work, was it all about one's own choices or some obscure forces playing their part in one's life? He questioned it when he and his fellow soldiers set foot in Cuba, back in the war, having had to face the mighty Spanish military, and he concluded back then that he had no choice but to enter a war dictated by other's interests. However if he hadn't happened to follow his father's career he wouldn't have been part of the army nor of the onslaught that followed. How about those caught in the cross fire? Did they have a choice? Or did the poor bastards just happen to be in the way of someone, like the outlaw Wade Johnson, who violated and took lives with no regard to anyone or anything? In any case Vincent accepted all the evilness done by man, even though it was hard to grasp at times, but now he was completely baffled by this turn of events, which was nowhere remotely near anything he knew or had experienced before. He had to wonder then, how much of a choice had he to end up in the "belly of the beast"? There were several opportunities to turn back, abandon this expedition and go back to the States, why didn't he? Perhaps because he wanted to prove to himself how much of a man he still was, after the dark years that he spent in the agony of intoxication and confusion, but that was not it. He did it all for the one reason, the only reason, which was giving him some comfort right at this moment, the fact that Gina and her children would benefit from the reward associated with this mission, assuming that the president kept to the deal. He had to, Vincent told himself, he had to, or else this ultimate punishment was in vain. He looked up towards the opening, he let the water drops come hard on his face, he mumbled a few curses, stood up once more, his heart heavy with regrets, but he also felt that he would not go into the darkness, not without a fight. He opened up his arms, clenched his fists and screamed as hard as his lungs would allow him. It echoed all around the darkened chamber many times over but got no response. However from that moment something told him that he was being watched and was perhaps not even alone. Fear took over his entire being once again, a fear so powerful that it almost paralysed him.

'What do you want from me,' he cried, 'where are you taking me,' he uttered, his voice hardly perceptible.

Instantly a blinding light hit his eyes so hard that he had to bring both his arms across his face, desperately trying to avoid it. He instinctively began to blink very rapidly as his vision was trying to adjust to the new demands of the environment. Moments later he let his hands down and looked around. Up in the air numerous silvery globes were floating, sending out red beams of light which covered his entire body. A screeching noise was accompanying the objects giving an even more frightening dimension to the entire scene. But it was about to get worse as within a few yards of him, there were another four beings, not human, small with bodies just about the size of a grown child, their skin grey in colour with big heads and large black eyes, tiny nostrils and small mouths. They looked like something between apes and humans with large maxillaries but small jaws, standing on two rather short legs, despite having long arms. But the one feature that stood out for Vincent was something that he felt he had in common with the strange creatures, just at that moment, and that was their facial expression, one of fear mixed with consternation. He did not have much time to further contemplate the bizarre scene because one of the many floating objects came close to him and began shooting energy beams, painful energy blasts that were hitting him simultaneously in various parts of his body, with the floating globe coming closer and closer each time. In a flash Vincent began walking backwards, turning his head all over the place in a desperate attempt to find cover. The floating orb suddenly sent out a strong beam of bright light, fixing on one of the many hexagonal openings on the wall. Instinctively Vincent ran to the edge of the water puddle, jumped over a small metallic wall that was containing the pool and began climbing up to the hexagonal hole that was lit by the powerful beam. As he reached the opening, the floating sphere, still very close to Vincent's back, stopped shooting out the painful lightning balls and started to make some sort of a metallic sound instead. Vincent looked back at it in pure dismay, only now realising that the entire exercise was to get him inside that pit. He knew now that yet more painful experiences were waiting at the bottom of that cavity. But was there any alternative? It seemed not, as the extended arm of the globe began forming yet another energy ball, getting ready to fire at any time. Vincent shook his head sideways in a sign that he had surrendered. In that instant the floating object retracted its' mighty weapon inside its own case, flew backwards a few yards and made a few burping like noises. Vincent put his legs in first and then slowly allowed his body to slide inside the moist walls of the cavity. With one last effort he glanced over to the chamber only to see the other creatures climbing up the wall following the exact same routine as he did. Some of them were screaming out loud the most unnatural holler that he ever heard, a high frequency howl that would sink right into a human's brain and make one's heart shrink in terror. As he took a couple of seconds to watch the horrifying show unfolding below, Vincent felt several spikes penetrating his body all over the place. This tepid fluid began invading his veins, warming up his entire being, giving him a rather pleasurable dizziness and unexpectedly making him not care a dime about anything. He felt some sort of gel right down at his feet which slowly started to fill up the gaps left in between his body and the cavity wall, while some sort of pipe entered his mouth, gradually advancing down into his throat until it divided into many smaller tubes that reached deep into his lungs. For whatever reason, this was not in any way painful and on the contrary Vincent felt as if the strange spikes and pipes were working for him, in some ways completing him. His awareness started to decrease with each second that passed until he could no longer keep up with anything. The very last thing he felt was that the gel coming from the bottom of the cavity covered his entire body, right up to his eyes and then Vincent went into blankness without any further warning, swallowed into the pit.

Chapter Thirteen

The city was just about coming back to life this late November morning. The air was crisp and felt fresh, not very cold but just right for this time of the year, besides it was looking like a sunny day ahead. Traffic was building up from the suburbs inbound and so the noise levels were mounting on both sides of the Potomac, around the metropolitan area. All sorts of people were pouring out from the subway stations, trains, cabs, personal cars, buses or whatever it was that brought them into the city. These were politicians, doctors, businessmen, builders, nurses, office workers even homeless people, all human nonetheless, with as many thoughts and ideas, personalities and goals, whatever these were. They gave a chaotic dimension to a scene that by its nature was not one of order, in spite of every one of these individuals having a specific purpose needing to be achieved, whatever that was. In a quieter area north side of the river, inside Fairfax County, in a rich establishment a phone was ringing. It kept going for minutes, then stopped and then rang again but it never went silent for more than a few seconds. Certainly it was not the only phone ringing, not in that area or elsewhere, but most of them rang just a couple of times and then stopped altogether. This one kept on going and the more it rang the more exasperating it became for the caller. That was because this was not just any call but an important call, as if someone's life depended on it. And maybe that was the case. Eventually this distant click indicating that someone was picking up, could be heard by the caller, and then a groan belonging to an old man, who sounded rather rough.

'Yes,' the voice said with much delay.

'There's been a glitch, an error of some sort that I need to inform you about,' the caller said.

'Is your line secure' the old man enquired casually.

'It's a brand new untraceable SIM card.'

'Good, you may elaborate then,' the old man demanded.

'It appeared to have been an exchange, all legit but then we realised it was from an unknown source...,' the caller explained almost fearful.

'Marcus, what the fuck are you talking about,' the old man shouted.

'Sir a human subject has been released; we then delivered one EBE out in Wisconsin and we shouldn't have.' Marcus concluded.

'What's the EBE's status?'

'Very much dead I am afraid.'

'I see, and the human?'

'Alive and well, but the location of his release does not match our coordinates. He was dropped somewhere in the rainforest in South America..., sir?'

'Aye, I hear you Marcus, this is very bad. Where is he now, the human, where is he and in whose custody?'

'Right here in Washington, in the military cuckoo yard, at Saint Elizabeth's.'

'Huh, what made you think it was a legitimate exchange?'

'Everything else seemed to be right, just the coordinates were wrong,' Marcus explained rashly.

'You're a damn fool Marcus, The Protectorate are not going to take this lightly, there will be consequences from it, harsh consequences, you hear?'

'I am well aware sir and willing to assume full responsibility for the entire glitch,' Marcus answered at once.

'No, you don't, I will speak to them. You make sure the human subject is eliminated and I will wash away any traces of your damn "glitch", God I hate that word, can you do that ahead of tonight's gathering?'

'But sir, allow me to say this at least, this human, he may be of unquantifiable value to us, to the Protectorate. I suspect he is coming from The Source, the ones we always suspected, from the TAG's.'

'Damn you Marcus, you follow the protocols now, this is a direct order,' the old man shouted, 'fuck the source, you don't know it. This thing gets out of hand you're dead, you hear,' he further exploded, adding more swearwords in the process.

'I remember you teaching me that any crisis is in fact an opportunity and frankly sir, I think this is a hell of an opportunity. Fuck the Protectorate, they're always searching for answers, the supreme truth as they call it, and when they get close to it they're too afraid to pull it off. I am a trustworthy member, and have done this for how long now? It doesn't even matter, but for long enough to know when such an opportunity crops up you don't just let it go.' Marcus said calmly.

'But you fucked up Marcus, many times, and it was me that had to clean your shit up. Now get your act together, do what you're supposed to be doing and contact me when it's done, preferably ahead of tonight, but if it's too risky just show up yourself, and say nothing about your fucking "glitch",' the old man replied before the line went dead. Marcus did not like that one bit and had to ask himself how much more of this shit would he have to put up with? He worked hard for The Protectorate and put his life on the line time and time again, what did he really do that for? Frankly speaking he didn't see much benefits for him. He agreed with their philosophy, he could see the point and he could see why their activity needed to be hidden, while the truth was to be protected at all costs, but he thought that the ultimate goal was always to find the special ones, The Ancient Gods, for they were the only ones to hold the "supreme truth". Now that he strongly felt that he was onto something, not just anything, but the strong probability that this particular subject had been released directly from The Source, the very people he served were telling him ignorantly to eliminate the subject, without even trying to at least explore the experience and the potential knowledge that the subject held. He realised how outrageous this kind of action would be, simply because this sort of thing never happened more than once in anyone's life time and even when and if it did happen you could never be sure of it, but for damn certain they needed to make an effort to try and find out. It was risky indeed, especially if the subject got into the wrong hands, which seemed to be case now, but it was worth trying nevertheless. Therefore just simply killing the subject was wrong. Marcus thought long and hard about this as he wandered the streets and felt strongly that for once he was going to go in a different direction to that in which The Protectorate wanted him to go. He knew that the penalty would be harsh; he knew that the people he was serving were no joke and they'd kill him or anyone without hesitation if they felt threatened in any way, but he also knew that he'd soon grow out of control and perhaps would be able to take them out, take over their assets and pursue the ultimate goal by himself. Perhaps that moment was now, considering that he had enough knowledge about the Protectorate and its members, something that they should have never allowed him to have. But he somehow earned their trust and even at times when he got it wrong, there was this old man to cover up and defend him within the Protectorate. One big mistake that was, from an organisation that had kept itself well underground for a very long time, only by not taking many false steps. This was something that Marcus admired on the one hand but on the other, he understood how this made the Protectorate become inflexible, even to the extent that they were now deviating from their original goal. Furthermore they seemed to have even forgotten the struggle that the Protectorate had to go through for such a long time, because in the past their access to various phenomena was limited, technology which allowed specific developments was very restricted and when all these came together, the Protectorate became suddenly more preoccupied with protecting itself rather than following the tasks. Well that was very wrong, Marcus decided, and he would change all that without delay, because delay meant losing a unique opportunity that might not show up again for another half century or more. With that in mind he flagged a cab and instructed the driver to take him to a specific address within the D.C. area.

On the other side of the Potomac River Dr Leary was inside his office. It was supposed to be just another ordinary day, business as usual sort of thing, but nothing had been quite usual in the past few days. Since Vincent Monroe landed in his back yard nothing seemed to be the same. Dr Leary had just heard one of the most amazing stories he had ever come across and he had to wonder whether there was any truth in it or whether the "story teller" was insane, far beyond anything he knew about psychiatric conditions. There were too many elements that were very well connected within Vincent's story, there were the strange circumstances in which he had been found, and there were hardly any signs of psychosis, not the kind of psychosis that Dr Leary has been working with every day for the past fifteen years anyway. There was clear evidence of some mental disturbance based rather on some severe life experience that left this man scarred, deeply traumatised indeed, but there again, if his story was real the effect of such dramatic events would damage someone's psyche in a very significant way. On the other hand Dr Palmer had another theory, according to which Vincent Monroe could have easily been an impostor, with a very rich imagination in which case he'd be classified as a pathological liar. She further argued, formulating yet another theory, that Vincent might have been part of some scientific expedition on the Amazon in more recent times, and gone through traumatic events and he might just have come across these belongings such as clothing items, note books, the backpack etc. But there was a problem with this particular assumption, for one it was farfetched Dr Leary found, and then to come across clothing items about a hundred years old, in good condition in the middle of the rainforest was close to impossible. However Dr Leary had to admit that, if he was to believe Vincent, which to some degree he did, the story told by Vincent was beyond farfetched, something that Dr Palmer strongly believed. Indeed it all sounded ridiculously bizarre, a mission to find "gods" that was directly sponsored by the White House about a hundred years ago, people facing painful deaths in the middle of the Amazon, pyramid-like structures, UFO's and one hell of a story of abduction? Dr Leary was just about to grab a copy book from his desk in order to go over the recordings, just in case he might find inaccuracies that needed to be noted, when he saw his cellular phone lighting up and vibrating frenetically on the desk. He picked it up anxiously as he recognised the number. It was the university lab where they dropped a few items from Vincent's belongings for analysis as agreed with Dr Palmer. He slowly brought the handset to his ear, took a deep breath and then pressed the answering key on the screen.

'Hello,' this young female voice began, 'is this Dr Leary?'

'Yes speaking,' he replied.

'Hi Dr Leary, I am Rachel from the college lab, just got the results back. You know those clothing items, DNA samples that your colleague dropped off a couple of days ago for testing and examination?!'

'Yes of course, what's the news then,' he enquired feeling his heart beat going up.

'Well I don't know if the results are going to shed any light onto your investigation or confuse you even further, but that's entirely up to you.' Rachel said, her voice betraying slight excitement.

'I am all ears, so fire away.' Dr Leary encouraged her.

'The clothing first of all, it proved to be over a hundred years old after the carbon fourteen testing but that's not all. We found traces of human hair that is as old, and it gets better,' she explained.

'Go on,' he said almost out of breath.

'We extracted DNA material from the hairs found in the clothing and when compared with the DNA sample you brought....., it's a perfect match.' Rachel stated.

'Can you e-mail these results to me ASAP, please? I think it sheds a hell of a light onto my investigation and for that matter I don't know how to thank you, but Rachel I appreciate it!' Dr Leary told her in a shaky voice.

'Consider it done. I am scanning the results as we speak. You should have them in a couple of minutes, and hey, don't mention it, always happy to help out. Have a nice day Adam,' she said, adding a personal note to the conversation.

After hanging up Adam could hardly control his emotions. He was over the moon to say the least. This new information was undoubtedly irrefutable no matter what concerns Dr Palmer was going to raise now. It was pure evidence that concurred with Vincent's story beyond a shadow of a doubt, a true case of alien abduction and not just any case of abduction, but something unheard of. However Dr Leary felt that he needed something more, so as to be convincing, not necessarily just for his colleague but others that would hear about this extraordinary find. He clearly recalled Vincent saying that he carried some sort of message, a message given to him by the "evil gods". He needed Vincent to reveal that message; whatever it was it must carry some weight and perhaps it would help elucidate any potential enigmas left in his case. There was one question that Dr Leary had no answer for, namely, this entire episode happened about one hundred years ago and yet Vincent was only about forty two years of age. Had time stood still for him from that time onwards? How was it possible to live over one hundred years without even ageing? It did not make any sense but Dr Leary knew someone, an old friend of his who happened to work in the field of theoretical physics and he might just hold the answer. Without delay he picked up his cell phone and quickly chose Mark Dowson from the contact list. It rang five times before someone answered.

'Mark, is that you?' Dr Leary asked sounding unsure as he did not recognised the voice.

'No, I will put my dad on in a sec,' this teenage voice said in a casual tone, 'may I ask who is calling?'

'Yes, I am a friend of your father, doctor Leary is my name, is that Andrew? Jeez it sounds like you've grown up!' Dr Leary said with enthusiasm, however he got no response from the teenage voice. Instead he could hear a grunt as if someone was clearing his voice.

'Adam, long time no see,' this man said and then grunted again, 'what made you call me after, what is it now, four years?'

'I need to sincerely apologise, Oh God time flies, doesn't it? How old is Andrew now?' Adam genuinely asked.

'Oh, he'll be fourteen next month; they just keep on growing..., now how are your family and all?'

'It's all good, look I promise to meet you for a drink soon and catch up, but now I need to ask you something that you may have an answer for, cause honestly I am baffled. I came across a man, he is a rather strange case, claims to have been lost somewhere in South America about a hundred years ago and then he was "taken up by some gods" as he describes it. Now, we know he is about forty two years old but...,

'Huh, is he for real, I mean how sick is this boy?' Mark asked almost laughing.

'Yeah, well I just don't know that yet, but I can't get into too much detail ..., um there are some things, some evidence that really supports his case,' Adam explained, a little self-consciously.

'Are you for real, like what evidence?' Mark asked, his tone switching from amusement to bewilderment.

'I can't really go into that but you know, some testing was carried out and the results are supportive. My question Mark is how is it possible not to age considering that, based on his story, these events took place that long ago?'

'Did you say that he was abducted, like ET's sort of things? I mean he was taken up there as you put it, is that like in space?'

'I think that is what he means alright, hard to tell, can you help me?'

'Well it's very simple, the answer I mean. If one goes on a journey and travels through space at the speed of light, let's say to a destination located fifty light years away from Earth and then decides to come back to Earth, then a hundred years would have passed here while for our traveller it all happened almost instantly. Do you read me?'

'Loud and clear, huh, it all sounds like a Luke Skywalker type of thing, Jeez this is nuts. Thanks Mark, promise to contact you a lot earlier than in four years' time, we'll catch up. Send my regards to Kimberly and to Andrew indeed!' Adam rushed to end the conversation.

'Will do, best of luck with your case, can't wait to learn more.' Mark said with a little disappointment in his voice.

'Mark, don't talk to anybody about this please, I just don't know where it will end up, you got to promise.' Adam explained realising he should perhaps have made the enquiry under a false pretext.

'My lips are sealed old buddy, see you soon then.' Mark assured him before getting off the line.

Being left to his own devices Dr Leary quickly accessed his e-mail, downloaded the files sent by Rachel and began printing them off. He had to acknowledge how overwhelming all this information was and how much pressure he was feeling. He had to share these findings with someone. He needed to talk to Dr Palmer. A moment later Dr Leary was running out onto the corridor feeling over excited. He went into his colleague's office without knocking, but to his disappointment Dr Palmer wasn't there. Coming out onto the corridor again he rushed down the stairs into the reception area but suddenly stopped in his tracks as he saw Dr Palmer talking to this man, who seemed rather strange. It wasn't necessarily the man's appearance but rather a gut feeling that

Dr Leary got by just looking at him. The man was tall, seemed young at perhaps a bit over thirty years of age, well-trimmed and wearing an expensive black suit, at first glance anyway. He wasn't just one of the ordinary people that would drop by, such as relatives or friends of the many patients in the hospital, and he looked more like a salesman or politician of some sort. Dr Leary decided not to interrupt whatever it was going on between the strange man and Dr Palmer. He returned to his office and called reception instead, asking that Dr Palmer should report to his office as soon as she was free. Minutes later Dr Kate Palmer showed up inside his office after politely knocking on the office door.

'Did you call for me,' she asked casually.

'Please take a seat, we need to talk.' he announced.

'Do we have anything new on Vincent?'

'Yes we do Kate; did you see him since coming in this morning?' Adam curiously enquired.

'I have, he is doing rather well I must say.'

'Considering..., like what?'

'Oh come on Adam, considering that hell of a story he just told us..., look I want to be on the same page, I just can't! I really want to believe him, but his story is utterly insane,' she justified.

'I need to see him again, today, I need to ask him about that message he claims to have been given. Hey look, the lab results are back, the ones from the college and they are shocking.'

'Wow that was quick because I am still waiting for his blood tests, only from across the yard. So what's the news?'

'Shocking news like I just said, um, here have a look for yourself!' Adam told her while handing her the printouts. He wanted to ask her about the man in the reception, rather badly, but then he thought that it may not be important at all since she didn't mention anything about her encounter.

'I don't really know what to say, are we really faced with alien abduction here? It's just that I'd never believed in such a thing, that it's just no more than some urban myth.' Kate articulated, sounding a little upset.

'It's what it is Kate! The results are speaking for themselves and right now we have nothing else, but I mean nothing to suggest otherwise.' Adam replied eagerly.

'What are we going to do now? Should we inform the government, perhaps NASA, or who?' she asked, never before having sounded so unsure.

'That I don't know yet. Let's call Vincent in, have a talk, and see what else he's got for us. I think that he needs protection and rest, like if it's all true we cannot even begin to think what his experience must have been like. I don't think that there are any written guidelines on how to deal with a case like this. Anyway I don't think we should expose Vincent, not now...., I simply don't know what's best.' Adam cried.

'You should perhaps consider what's best for everybody including ourselves, but whatever you decide I will back you up, just so you know. I will go and get Vincent and bring some coffee while I am at it, I am sure you'd want one.' Kate told him as she made her way out.

'Sweet and strong please, no milk,' he shouted after her.

Minutes later a fresh coffee aroma was filling the room, which, mixed with the scents of masculine cologne and fine ladies' perfume gave quite an aristocratic dimension to an office that was filled with tasteful furniture items. Dr Leary was rather conservative when it came to taste, but very open minded otherwise. He was sitting on his luxurious armchair appearing rather relaxed, in spite of the pressure he was under considering that right at that moment he had no answer to what he would do about Vincent's case. Moreover he was eager to ask Vincent more questions and perhaps, depending on the answers, they would have a better picture with regard to Vincent's future. He meticulously took his china coffee cup, took a sip from the aromatic black liquid and then sat the cup back on the saucer.

'Mr Monroe,' he said with an air of excitement in his voice, 'please have a seat and some coffee if you so wish.'

'Thank you, I sense that you perhaps made up your mind regarding my situation, am I right?' Vincent began without hesitation.

'In a way, yes we did, but it is vital that you tell us a little more, maybe about that message you claim to have been given by those "gods".' Adam replied informally.

'What difference does it make to you?' Vincent asked curiously.

'Let me tell you Vincent,' Adam said getting up from his seat, 'you are it, you are who you claim to be, we concluded that based on the tests we carried out, and we can't say any different unless, on the off chance, it is proven otherwise. But having said that there are still questions that need answers, do you understand so far?'

'I am not very sure I do doctor, on the one hand you are telling me that you have some sort of proof of my identity but then you also say that it is still unclear, unless I give more answers. What more do you need to know?' Vincent asked, his voice a little shaky.

'Okay we have the results like I said, which show that you tell the truth and frankly I believe you and I think you have convinced Dr Palmer as well. Unfortunately Vincent we live in a world where stories such as this are not easily accepted. If this story goes outside these four walls people will ask the most awkward questions, they will ridicule you and they will ridicule us which may also mean that our careers as doctors will end. Therefore we need further strong evidence that you are telling the truth and you have been indeed abducted by extraterrestrials, by the way that is how we call these "gods" of yours, meaning that they are not from this world but coming from elsewhere.' Adam explained.

'Why do they need to know? I mean apart from maybe the President, who else needs to know my story and why?' Vincent enquired, sounding even more confused.

'Because that is how it works. Your story is unique and honestly we are very unsure of how to go about it, who to tell and what consequences it may imply for you, if we do tell anyone. Our responsibility as your doctors is also to protect you and make sure that no harm comes your way, in any shape or form, so please help us out!' Adam pleaded.

'I understand that my story sounds weird, I get it. What I don't get is why you need to protect me and even yourselves, protect from what? I mean no harm, but ask me more and I will answer, if that's what needs to be done. I am just rather aware that the message you're asking me to share may not bring more light into this, and on the contrary it may bring more shadow. It's not as if they handed me a letter, it's all in my head and it's all numbers that make no sense to me...., I have no words in the message doctor.' Vincent explained.

'Can you reproduce it, the message can you tell it?' Dr Palmer enquired after her moments of silence.

'I see it, it flickers in my head when I close my eyes,' he told her in a low voice. Without waiting Vincent grabbed a small notebook and a pencil from the desk and started scribbling numbers, only zeros and ones, randomly filling page after page about six in total. The whole spectacle was weird to watch, as if Vincent went into some sort of a trance, keeping his eyes closed the whole time while his hand was doing the work with no apparent coordination from the brain. Both psychiatrists were baffled once again by this development. When done Vincent sat the pen down beside the notebook, slightly shook his head and then opened his eyes, not even looking at the numbers he had just written.

'Are you satisfied doctors,' he asked in a mysterious tone.

'Hey Vincent, I am so sorry to have put you through this and we can't even imagine what you may have been through, but look at our position here, like I already explained your case is unique, nothing close to what we have dealt with before, but thank you for having the courage to go all the way!' Adam told him apologetically.

'Aye, no harm done, I don't know what it means but that is it. They somehow put that inside my head.'

'I know someone who can help make sense of it, it's called a binary code. It is like a language those "knowledge machines", as you call them, use in order to work.' Dr Palmer told him as she softly landed her hand on his shoulder. Vincent looked up at her and nodded in a sign that he'd like to leave. He then got up and walked to the door without looking back at the two doctors, however, he said "thank you" on his way out.

'Poor fellow,' Kate cried as the door closed shut behind Vincent. 'I think I should take this to someone I know, he works as a software developer, he'd know about binary codes and stuff,' she said informally.

'Do you trust him? I mean this computer guy with the message, God knows what it says.' Adam told her hastily.

'Of course I do, it's an old friend besides I am not going to tell him who it's from or anything,' she replied, sounding a little irritated.

'Yeah alright, but what if the message reveals something very important? Your friend may become eager to find out what it is. Look Kate I don't mean to sound paranoid but this is big and we need to be careful who we share this information with. By the way I've seen you talking to some stranger down in the reception area, who was he?' Adam quickly added.

'Oh my God Adam, if don't mean to be paranoid you are certainly acting like it, are you spying on me?' Kate asked, adding annoyance to the confusion in her voice.

'I saw that accidently and I thought I'd ask. I just need you to comprehend the complexity of our case, that's all,' he said casually.

'Yeah I get it, but regarding many aspects of this matter, our knowledge is rather limited. Do you know anything about binary codes, because you can feel free to translate all that shit! And the man in the hall was just an insurance guy trying to sell a new product to our patients.' Kate explained feeling a little embarrassed by her colleague's attitude.

'Did you check him out,' he asked bluntly.

'I got a name, Marcus something, didn't need to go into details, I just sent him out the way he came in. I told him that our clients are insured by the government.' Kate insisted, with annoyance.

'That's exactly it; he should have known about this hospital before coming in trying to sell health insurance to people already insured. Kate please listen to me, it doesn't matter how crazy I may sound or act but we have a serious matter on our hands. I don't know who to talk to about any of this, but surely we need to trust that person with our lives.' Adam pleaded.

'Alright, okay but I can't just assume that any visitor coming into that entrance is potentially looking for Vincent, besides who else knows about his story?'

'Well the doctor who found him in Brazil knows about him anyway, Dr Wilson I think,' he told her visibly pretending not to recall her name. In fact he knew Linda Wilson very well.

'We should call her you know, just to ask her for a more detailed description of how they found Vincent.' Kate suggested.

'Um! Don't know about that but I will consider it when I'm writing the report, the question is, for whom am I to write the report? I don't think the Department of Health would want to read it. What do we do with this Kate? Where do we go from here? These are the questions I need answers for, like right now.' Adam explained genuinely.

'I will get the code translated; I will scan it and send it to my friend. We might have it back just after lunch, then we can weight up our options.' Dr Palmer concluded.

'I agree. You do that while I run the recordings once more, see what I've missed if anything,' he said.

Left on his own Adam felt no motivation to run the recordings anymore in spite of having promised to do so. He considered it to be pointless since Vincent's story concurred with indubitable evidence. Instead he drifted deep into thought, attempting to find a way forward, knowing that options were few and far between, especially if they were to try to limit any sort of negative consequences to Vincent's wellbeing as well as to their professional careers. He knew that a report needed to be drafted and based on the evidence, conclusions from such report would be rather simple in as far as Vincent Monroe was telling the truth and in spite of severe trauma suffered by the client as a result of his bizarre experience, the man was not hallucinating or making this up. This was something that Adam could confirm based on his professional opinion, without any hesitation. It was then up to the commission examining the report to come up with a service provision that should at least attempt to serve Vincent's needs, and to Adam that was a problem indeed. What service would that be, considering that no official or recognised clinic or agency was specialised in dealing with what he'd call a genuine alien abduction case. He knew of private consultants and therapists who were offering services to people who claimed to have been abducted but who had little or no evidence for their claims. Vincent was just different, his case was strong and by no means was it lacking proof. What would they do with him then? What sort of future would lay ahead for this man, considering his very disturbing but truthful past? It was hard to tell and in spite of having the responsibility to protect Vincent's best interests, Adam knew that the case would soon be out of his hands. He concluded that once he had written the report and handed it over to the commission it would be entirely up to the Department to deal with the case, whether they liked it or not. The only thing that Adam and his colleague could do was to write a clear report and make recommendations that were not biased, and also to try and prepare Vincent to face prejudice and even unfair judgement in whatever form that was to come. Feeling somewhat at ease with his own thoughts Adam went back to his desk and quickly began typing the draft for his report only to be interrupted by the sudden entrance of his colleague. She came in holding an A4 piece of paper, but looking rather disappointed. She laid it on the desk right under Adam's eyes who quickly understood that he was looking at a printed version of an e-mail. He read it twice and hardly believed his eyes at what he saw.

'Is this it, for Christ sake, is this it,' he shouted in desperation.

'I had the same reaction reading it, but it's what it is.' Kate responded in a low voice.

"Search inside my head you must for there is more" was inscribed in bold capital letters underneath the common enough text of the e-mail. Adam kept on reading the sentence out loud, over and over again, as if attempting to find more but there was nothing else but those few words.

'What the hell is this supposes to mean, search inside my head, like how? Is this saying that perhaps we should go down the road of hypnosis or what,' he asked in pure bewilderment.

'I don't know, I am as lost as you are, honestly haven't got the faintest idea of how we are to further proceed with the case. Did you have any more thoughts on it?' Kate enquired nervously.

'I am just going to type up my report,' Adam said while shaking his head, 'that's it Kate, we write the report based on pure observation and evidence, not in any way biased and make further recommendations about his future care. We need to tell Vincent what could be ahead of him, after we hand over that report, and he will likely move to a long term care facility,' he concluded.

'I could not agree more, it is up to the Department, we did whatever it was in our power to do as professionals, to help this guy, but his case is way out of anything we've done in this clinic,' she said sounding a little more optimistic.

Chapter Fourteen

He was trying to remember when last he felt so disgusted by anything at all, but nothing really came to mind. It didn't because deceiving, lying and even killing were no strange territory, not at all, not to Marcus. He was groomed in a cold blooded spirit, raised to seek and destroy and taught to sacrifice anything or anyone in search for the "supreme truth". He was trained exceptionally in the military arts and never allowed to express weakness, neither was he taught love nor was he allowed to feel empathy. Instead he had vast knowledge about pretty much everything there was to know from politics to society, from global economy to astronomy and beyond. He was also in possession of a large variety of skills, knowing when and how to use each of these skills, from being charming and delicate to being ruthless and cruel, but above all he had a whole load of information, classified top secret information; the kind that gets people killed or enslaved no matter who these people are. And right at this moment there were eleven bodies being incinerated right in front of his eyes. It was the smell of the burning flesh and fat that was making Marcus feel disgusted. But it was done and now he was the only member from "The Order of Osiris and Seth", also calling themselves "The Protectorate", that was still standing, feeling more alive than ever and having the prospect of a very bright future indeed. Because once he had them out of the way forever it was much easier to follow the path to The Source, especially now when The Source appeared to have released a subject of paramount importance. He attempted to convince the other members that he now had enough intelligence on the subject, more or less direct proof that this particular individual was released by The Source, but they wouldn't have it. They would not even consider it, even after Marcus had played a recorded conversation between the two doctors in whose care the subject was, as Marcus carefully placed a sophisticated recording device on one of the doctors, earlier that day. He really thought that the material would be sufficient to convince the other members of the potential authenticity of the case, but they couldn't see it, which led Marcus to make this extreme choice, eliminate the organisation and take fate in his own hands. He was now pleased with the choice after realising that The Order had become obsolete and worthless in spite of its tumultuous past. It had gone through so many stages of development in over three centuries of existence. Initially The Order had been part of a larger organisation; it then broke up, rapidly becoming very obscure. At times it had been even militarised, after the Roswell incident, however it managed to emerge free, eventually, from any military or political influence. Its active members were either ex secret services, "the elders", or else people directly employed by them, "the youth", Marcus amongst them. These were people who could sway and manipulate anything and anyone at all levels and yet none of the individuals that Marcus had just killed even existed in any official records, as real persons with real identities, but they were rather "shadows" after erasing their past without a trace. In fact they were never recorded in any public annals and just took different identities which they either stole or simply made up. As an organisation it had always obtained large amounts of money via direct access to governmental funds worldwide or through sponsors, some of the richest people in the world, out of whom some were willingly sponsoring The Order, the so called "inactive members", while others were not even aware of allowing large amounts of capital to be used by The Order. These were called the "investors". During its long existence The Order has been involved in all sorts of things such as political putsches, economic espionage, financial fraud, conspiracy and much more, activities that also generated large amounts of money, access to very sensitive information and many other benefits. The Order ensured its continuity by always passing its legacy to new recruits, ever since it came into existence. There were always twelve, six elders and six juniors, but now there was only him, the last man standing and the sole heir to The Order's legacy which was vast and sophisticated, but so was he, because now Marcus was The Order. Without a shadow of remorse he only focused on his tasks ahead. He needed direct access to the subject and it appeared that the two psychiatrists were in his way, however he needed to be very smart in how he acted from this moment forward. Ironically it was a lot easier to eliminate people that did "not exist" than people with real identities, nonetheless he would not hesitate to take them out too. With that in mind Marcus walked out from the burning building, and walked through dark alleyways until he made himself unseen, leaving behind the raging blaze and many sirens rushing to the site. It didn't bother him as there was nothing that would be found by any forensic investigators that would link him to the site, he didn't "exist" either. He was only rushing so that he would be on time for his next "appointment", likely involving yet another killing. The night was still young, so he grabbed a cab as soon as he reached the residential area nearby and asked the driver to take him to an address just outside the metropolitan area. Once there Marcus paid the driver and got out still having to walk about a mile before reaching his real destination. He stopped just outside this eight storey apartment block, spent a few minutes contemplating it and then entered the underground car park after carefully avoiding the CCTV system. He walked among the many cars and noticed that the one he was looking for wasn't there just yet, as he expected, so he stepped out of sight into a more obscure corner. It wasn't long before a gold Chevrolet Sonic pulled into the space reserved for apartment 18. A middle aged, nicely shaped woman got out of it holding a laptop case in her left hand. She took a few steps away from the car as the sounds and flashing indicator lights of the central locking device disturbed the serene peace of the place. It was in that moment that Marcus stepped out of his dark corner coming up right behind her, still unnoticed.

'Dr Palmer,' he asked casually. Instead of a reply Kate whined in such way that she betrayed both her fear and surprise, stopped walking and slowly spun around to face the caller.

'Marcus,' she managed to whisper appearing even more frightened than a second ago. 'What are you....., I mean you are not selling health insurance are you?'

'No, I'm not but I am a seller nonetheless.'

'...and what is it that you sell,' asked Kate her voice fading away.

'Mostly death, but it is important in what form the buyer wants it, sluggish and agonizing or swift and painless. Now you too have that choice, it just depends on a few answers I need. So just picture this, I can drag you upstairs, rape you and make an absolute mess having you bleed all over your floors, be in pain for hours, I don't have a problem with that. Alternatively I can ask you to get back into your car and then I can swiftly shoot you from the back of your head, just a couple of little holes for the forensic guys to measure, no pain,' said Marcus, sounding very amused.

'I don't know if I can help you at all, I don't understand what would interest you so much, I am just a psychiatrist,' said Kate genuinely confused.

'Oh, but you can! Subject Vincent, tell me about him. All the results confirm that he is a genuine case of abduction isn't he?'

'It would appear so, but...,'

'He is unique, now what did he tell you about the "gods" he saw, did he describe them?'

'He said that they were evil, they've done things to him, bad things.'

'Um, I see! It would appear that he comes from The Source. You see, my forefathers and I were chasing The Source for a long time and by God has it been hard to get close? Yes it has, but now I am so close to it. And I can't allow you to live, as surely you and your colleague will potentially compromise my mission; therefore I have no choice but to eliminate the two of you. Before I do that I need to ask you one more question, and be careful, this is the one million dollar one, who found him, who found Vincent in the rainforest?'

'The military I believe, it was the military who found him,' said Kate with little conviction as she was lying.

'Oh, surely I expected you to come up with a name. You see my Intel on the matter does not concur with your answer. I know the military brought him to the States but they didn't find him. It was someone else, don't worry, I will find out from Dr Leary....'

'It was a local tribe who found him and I don't know how he got to our military, but that's all I know, I swear,' shouted Kate trying to sound convincing.

'That's all I know, I swear,' Marcus imitated her voice in a sarcastic way, 'you lying bitch, you're lucky that I am on a very tight schedule, so get into your fucking car and stop moaning,' he ordered her severely.

She knew that it was the end. She knew that there was no way of fighting him, and pissing him off even more would likely attract painful consequences. The only choice she had was to die with a little dignity, although this was no ordinary killing but an execution and there was no dignity to be claimed here. She got into her car, sat in the driver's seat as instructed, felt his breath increasing at the back of her head, he even played with her hair before sticking something metallic and cold right at the top of her neck and then this flash of light struck her eyes followed by nothing. Just another lifeless body with no story to tell would be found early in the morning for there would be no connection between Kate and her killer.

She would now become just a name in a file amongst countless others, thrown inside some box marked "unsolved".

Later that night, in a different area of the city, tiredness was taking its toll, slowly beginning to claim Adam's entire being. He'd been typing his report on Vincent for most of the evening with hardly any break and it was not only the work itself that was tremendous in any case, but this report was so different from any other that he had written. He needed to choose his words very carefully so that there was no room for interpretation by any members of the commission, which could in essence become damaging to the whole case. It would arguably be considered by the commission that this case might be a hoax and one could not blame them, as it is not every day that they would come face to face with a genuine alien abduction case. The report so far was looking good, Adam decided, the fact that they picked an external lab to carry out essential tests was a plus beyond doubt. There again these folks were not the easiest to convince and likely they would run their own investigation, but so be it, they could not possibly reach a different conclusion. Vincent's case was what it was, and no one could say otherwise unless they were to find new evidence, strong enough as to demolish his entire case. The only thing that Adam felt let down by was the translation of the binary code message. He thought of it as a disappointment as he had expected something else entirely, couldn't name exactly what, but had to accept what the message said. And perhaps there was meaning in those few words, just that he could not find it as of yet. He looked over at the blank page on the computer screen, the page on which he was supposed to write his conclusions, but he realised that it would be better if he returned to it in the morning, with a refreshed and rested mind, as it would be one of the most important pieces of the report. He saved the document and switched off the Mac Pro notebook before leaving the office. As he walked outside towards the car park he became restless and for no apparent reason he began to feel fearful. He instantly started to think of his colleague and her earlier encounter with the stranger in the black expensive suit. Trying to sell health insurance in a state facility was something no one would do unless they were dumb, but that guy did not look dumb at all; on the contrary he looked very smart and very focused. Indeed Adam only had a quick glance at the man but still, he sensed something rather odd when he saw him. He drove home that night unable to push his thoughts away, but recognising that his restlessness might be generated by this complicated case. He fought hard to put the matter to rest, but his anxiety increased as he could hear all these sirens in the far distance. There was always something happening in D.C., no matter what time of the day or night, but some crime or accident, something bizarre was always taking place, with the accuracy of a Swiss watch. The Capital of the most powerful country in the world was for most an absolute living inferno, one of the least safe cities in the entire western hemisphere. When he got up the next morning Adam followed the usual routine, having coffee with his wife, seeing the kids off to school and reading the paper. His attention was drawn to an article according to which a fire had broken out in an abandoned warehouse in an industrial area of Fairfax County, where eleven bodies were found but not identified. Further on the authorities were treating the incident as suspicious. Adam didn't make much of it, naturally it wasn't much of a concern to him, but he found it odd that the fire had broken out in a vacant warehouse killing eleven people. What were they doing there? It would be unusual to find eleven people in one place, where it just so happened that a fire had broken out, killing all of them. Suspicious to say the least, Adam concluded, just as he got into his car. Back in his office he called for Dr Palmer only to learn that she had not yet arrived. It was now coming up to 10 AM and Kate wasn't known for lateness, not without calling in anyway. His anxiety from the previous night came back with a vengeance, more like a gut instinct that something bad had happened and if it hadn't, it was about to. He called Kate on her cellular but it went straight into voice mail. That too was unusual, knowing Kate she'd always answer her phone, day or night. He thought to ring the police department and announce that she was missing but something killed that idea straight off. Instead Adam ran out to his car and drove over to her place, only to come up to the junction before her apartment building and see that the area was sealed off, while a few squad cars and one ambulance were stationed around the underground parking entrance of the apartment block. He indicated left and drove away from the area knowing that his colleague was no longer amongst the living. Only one thing came into Adam's mind, to try and move Vincent out of the clinic for he was no longer safe. He also knew that whoever that man was, the one that met Kate in the reception area the previous day, he was after Vincent and likely he had now claimed Kate's life. It all sounded insane, he admitted, but it was too much of a coincidence when he put everything together. He drove for a couple of blocks and pulled over at the first phone booth he saw, took his cellular phone and searched for Linda Wilson in his contacts. He quickly dialled the number after throwing a quarter in the coin slot and waited for a good few seconds. He couldn't say why he called her, or even if she was back in D.C. but he could not think of anyone else. He needed someone to talk to about these events and that someone could not be any of the official people. Perhaps he was overreacting but Adam always put value on the principal that says "better safe than sorry". It so happened that he and Linda got very close while in medical school at Harvard, even had a short-lived relationship, but then managed to stay friends forever after. He was surprised but equally excited when Vincent showed up at his clinic, to see that the paper work was signed by Linda Wilson. Adam felt that it might be a way of reconnecting with a dear friend with whom he had lost contact in the past while. So why was he calling her now? Was it purely from sentimental considerations or was it a professional reason? Adam concluded that it was neither, but rather desperation as he needed someone to help him to help Vincent. Linda's phone kept ringing, which was a good sign, meaning that she was in the country otherwise the call would be forwarded to voicemail. When she eventually picked up she sounded as if she had been sleeping, her voice melodious nevertheless.

'Hello! Who is this,' she asked.

'It's Adam, Adam Leary and I need to talk to you, sorry sounds like I've woken you,' cried Adam.

'Oh God Adam, that's a blast from the past, yeah as a matter of fact you did wake me, so must be important.'

'Yeah, blast from the past, it's just that I need your help, can you meet me in the old café near my clinic, please it's important,' said Adam quickly.

'It sounds like it, are you alright?'

'Yeah, for now, look it's about that fellow you found in Brazil, I can't say much over the phone, just meet me in an hour!'

'Okay, just landed here yesterday, had a lot in mind for today. I would have called you to enquire about him anyway, how is he? Oh boy was that a strange find?'

'It got even stranger and now I feel he is in danger, but please no questions now, I will explain everything when I see you. One hour, the old cafe, please be there,' said Adam hanging up, not giving her a chance to reply. He wanted to sound a bit dramatic so that she would have no choice but to come. Whether it was fair or not was a different matter but it was certainly just. One hour later Dr Leary found himself in the cafe near St Elizabeth's, after leaving his car many blocks away from the area. His cell phone rang time and time again which he kept on ignoring. It was the clinic each time and when he listened to his voice messages he learnt that two detectives from the PD were insisting that he met up with them. The last message informed Adam that the detectives were to return after lunch time. He ordered coffee after placing his phone back into his pocket and took a newspaper from the counter before picking a table farther away from the entrance. There were no other customers at this time of the day. The front page of The Post had a picture with the incinerated warehouse which he had seen earlier this morning in a different paper. More pictures on the second page showed eleven black plastic bags of different sizes aligned in a row. Adam did not bother to read the rather large editorial beneath the pictures; he already knew what that was about. A few minutes later Linda Wilson entered the cafe, dressed casually but looking well nonetheless, she always did. She ordered coffee as well before coming to sit down at his table.

'So what's this about,' she asked going straight to the point.

'I am going to be short, time is not on my side, so if you just listen to what I say with no interruptions it would be great, keep any questions for when I am done, although I can't tell you a lot at this time, but what I am going to say is no exaggeration or fabrication of mine. Are we clear,' said Adam informally. She nodded and made a gesture with her hand as if to suggest that her lips are sealed. For the next twenty minutes Linda just listened to what seemed to be a mad story from some mad film, but certainly believed it, one because she knew Adam way too well, but also Vincent's story began with her only a short while ago, back in the rainforest. However the last part where Adam's colleague was presumably dead and some stranger in a black expensive suit was after Vincent, sounded a little farfetched.

'I don't really know what to say Adam! You don't have proof to associate Dr Palmer's death with any of this, if she is dead at all,' said Linda circumspectly.

'You've got to trust me on this! It's not one of these situations where you find evidence all over the place. Just think, this man Vincent, he is unique and then all these strange happenings in the past twenty-four hours, I know it sounds mad but these things are related,' he cried.

'What do you want me to do?' I don't see how I could be of help.' Linda explained.

'I need you to take Vincent into your care for a little while and I know that's a lot to ask but I don't know who else to go to! Just for a few days until I figure out what to do, to see who

I can speak with in the Department and I will take it from there,' he pleaded.

'Are you out of your mind? I can't take care of him..., I have my daughter staying with me, besides you can't just sneak out a patient, I assume he is registered and all!'

'No one knows his story..., I will make the paper work. I will see to it that a family member claims him or something! Please I am afraid for his life, and even mine, however if he is secure I can handle the rest. It's just a few days, I beg you!' Adam insisted.

'This is insane, not to mention highly illegal and unprofessional. Um, never mind, I will do it, don't ask why but somehow I trust you..., what's the plan?'

'Give it about twenty minutes, go to the south entrance and identify yourself with your driving licence. I will bring Vincent over and, huh, by the way don't use your cell phone to call me. I will call you from public phones. Thanks,' said Adam getting ready to go.

'Hold on,' Linda said reaching for her handbag and pulling out a small plastic bag containing a sample of some weird stuff, 'this peeled off Vincent's skin, when we found him, it's like a dry gel, it's strange because I ran a couple of tests back in Brazil and it appears to be organic but the composition is fucked up, nothing like I've seen before,' she explained.

'I will put that along with other evidence we have on his case, thanks! I will see you shortly.' said Adam while grabbing the small bag from her hand.

Chapter Fifteen

It had been days since Vincent came to stay with Dr Wilson, way outside the D.C. area, and way outside any kind of modern civilisation at all. This was a cabin somewhere in Virginia, up in the Blue Ridge Mountains in the middle of a forest, quiet and beautiful, a leap back in time as it appeared, which did a world of good for Vincent. It was here for the past few days that Vincent began to regain his health and his thoughts, his strength and his will, but above all he began to accept what it was that had happened to him. It was important that he started to come to terms with the fact that he left his world and inexplicably he came back to life in this world, separated by a century of unknown, of pure mystery that only he could possibly recall. After he was taken up into the cosmos, on what people now call a spacecraft -not so far from his own term of "sky boat"- he could hardly remember what had happened, just some faint memories of strange places, things, noises and emotions. These were coming back to him more and more in his sleep at night, like short fragments of nightmares, often waking him up in sweat and tears. Almost every time Linda would be close by, trying to reassure him that it was all a bad dream, but it wasn't. These images and emotions belonged somewhere in time, a different time indeed, but part of his past experiences nevertheless. If he was to try and define what exactly it was that he had experienced it would likely go along the lines of suffering and pain, violent undertakings and death, he saw death embracing many beings amongst which were some of the "evil gods" themselves. The challenge Vincent faced trying to explain any of these episodes, which he certainly witnessed, was his own understanding of it, which was lacking, simply because he could not see why he had been taken there to witness it all. Accepting what had happened to him was surely a step forward to recovery, or so it was explained, understanding it was just close to impossible. Dr Leary told him that a commission of doctors from different areas of medicine were to evaluate his very complex case and come up with ways to help him deal with both the physical and psychological traumas. That was the last thing he heard from him as he left the clinic and he hadn't had any sign from the psychiatrist since. For that reason Ms Wilson seemed very concerned, but she wouldn't share much and Vincent could sense that something was not quite right. It was even more difficult for him to make sense of this world, to see how things worked and he felt disempowered because he didn't belong here, not in this space nor in this time. But in spite of all this he knew very well that something had gone wrong, the minute he was told that he needed to leave the clinic. He wasn't given much of a reason either, except that it would be good if he spent a few days outside the hospital which he had to admit was at least desirable. But he also felt that it wasn't the reason and in addition he could read dread in Dr Leary's eyes and now there was no sign of him whatsoever. In any case his choices were between limited and non-existent, something that made him feel very uncomfortable, but he had to accept this fate. On the upside he enjoyed the place and the company of Linda and her daughter Tess, an energetic and funny youngster who was listening to some weird noise which she called heavy metal music on a handheld gadget that in spite of its' small size was quite loud. She became very attached to Vincent, enjoying spending time with him fishing in the lake nearby. She jokingly started to call him "My Amaranthine" , based on a song she liked, the only song that sounded near enough melodious, as for the rest it was just bad noise. Tess didn't know his entire story but her mother had told her that Vincent had got lost in the rainforest while on an expedition and luckily the tribe which Linda was working for had found him. That in itself was quite a story for a teenager who was avid for adventure and romance, and who since then had begun seeing him as some kind of hero. In spite of his rather precarious and very uncertain situation, Vincent would often lose himself in the moment and benefit from the good vibes generated by both his kind companions and the serenity of the place - well when the weird music wasn't being played. So alternating between pleasant and painful moments became the everyday routine for Vincent but he knew well it wasn't going to be for much longer. The agonising fact was that he didn't know what would follow, and part of him would want this moment to last for a lifetime. Having much time to reflect upon things he'd often wonder what would have happened to him if he hadn't gone on the expedition at all. Perhaps it wouldn't be that hard to figure out, he would likely have consumed the remainder of his days intoxicated and unaware of the time passing him by with no purpose. But then what about Gina and her kids, what had happened to them since he left, would there be any way finding out? It seemed one could find out anything one wanted in this bizarre world where information pops out instantly on "knowledge machines" of all kinds. This curiosity begun to grow stronger and stronger with each passing moment until one night he decided to approach Linda and speak what was on his mind. It was late, way after Tess had gone to bed when he found Linda sitting quietly on the front porch with a glass of wine, seemingly contemplating the stars.

'Quite a big place, the sky' he whispered as he came close, 'I can't help but ask how far did they take me,' he added.

'Is that relevant at all Vincent? It just makes me realise that in spite of its beauty it's rather a scary place, way too vast and way too dark,' she replied peacefully.

'It is to me, but you may be right, it's less significant where or how far when you set that against the one question that grinds me down, eats me from inside out. Why!?'

'I am sorry Vincent, but perhaps you will never find an answer and I can only imagine how it may hurt, but if you knew, would you be better or worse off right at this moment,' asked Linda placidly.

'Um, that's the thing I can't really tell, but without knowing it's as if I am left blind, left in the darkness,' he cried.

'I can understand that, someone will help once the commission looks at your case, they will find a way for you to deal with the pain,' she explained.

'I take it that there is no word from Dr Leary, is he alright I wonder?'

'Yeah, I am a little concerned too. I will call him first thing in the morning, we need to move on.'

'I need to thank you for your kindness, it's been very nice to spend this time here with yourself and Tess,' said Vincent before turning his eyes back to the clear sky. 'Did man get to go up there yet, as far as I can see much progress has been made, but did we go up into the skies,' he enquired curiously.

'Yes, they did, they went to the Moon a few times and then they sent space probes, machines with no people on board, farther away. Look I am not very up to date with any of that but Tess is. She is mad about these things so I am sure she'll be eager to answer all these questions for you. She wants to be an astronomer you know?' Linda told him, betraying a sense of pride.

'She is a cute kid, you are truly blessed with her,' he remarked. 'Not that it is any of my affair but you never mentioned Tess's father,' he said purposely, so that he could tell her what he had on his mind.

'Huh,' she sighed, 'what's to tell? He is a well-known politician in Washington. We've been married for about ten years and then I found out that he was having a number of affairs with different women so I decided to separate. It was hard on Tess but would have been worse if we had stayed together. We were fighting a lot, so it's better now. She gets to stay with him while I am away and vice versa. He is a good man and really loves her. That was never a problem. How about you Vincent, did you have anyone before departing on your expedition?'

'I cared about someone alright, only I was drunk a lot. By God do I regret that now? She, Gina was her name, was a good woman, took me into her home and looked after me hoping that I would change but I couldn't. After the horrors I had seen in the war I could no longer find peace, so it was just the whiskey that gave me comfort..., uh, I thought I could get better, my wounds would heal, but the more I drank the deeper I went into a state of confusion. She had two children Gina did, so she had to put them before me. She asked me to move out, to sleep in her barn so that the children wouldn't see me constantly intoxicated. It was the expedition that sobered me up, but way too late though after leaving her behind, her and the children. The president offered the reward of twenty thousand dollars in cash. I went on that damn expedition so that the reward would go to Gina and the young ones, to fix up the ranch and send her children to school, for I was no help to them but only a nuisance up to that day, so I was happy then knowing that something good was coming from this...., I am so sorry,' he said sobbing almost, 'do you think using those gadgets, um the ones that seem to know about people and stuff, do you think we can find out what happened to her,' Vincent cried out loud.

Linda felt moved by this story, so much so that she began shaking. She nodded while stepping towards him and then hugging him with compassion not knowing how else to react.

'We can try! We can look inside to see if the "gadget" knows about your Gina, but I can't promise much,' she said almost crying as well.

She then spent a few seconds in his arms and brought her right hand around Vincent's neck as she softly kissed him on the cheek. It was at that moment that she felt something odd at the tips of her fingers. It was something solid and pointed that was sticking out behind Vincent's left ear, something that should not be there. She stepped back and looked at him with dismay, without intending to panic him. He instinctively brought his left hand up and touched that same area. Within a moment he went down on his knees as if something hard had hit him in the back of his head, his eyes went blank and although they were fixed on Linda, it felt like he was looking through her and not at her. A slight tremor could be seen on his lips seeming to communicate something, but with no sounds. It was rather terrifying even for an experienced doctor to see that whole scene which left her stunned for a short while. She then kneeled right in front of him calling his name, but got nothing near a reply, nothing else but a groan. Unexpectedly Vincent just appeared to come back to his senses as he suddenly started to question what he was doing on his knees. Tess came out onto the porch too, after being woken up by her mother's loud voice.

'What's wrong mom,' she shouted in a panicky tone.

'I don't know it's Vincent, he had some sort of a seizure but seems okay now, go back to bed honey, it's alright,' said Linda not taking her eyes of Vincent. She then rushed to help him stand up and even to walk a few yards back inside the house.

'What were we doing out there at this time of night, by God is cold,' he asked seemingly unaware of anything prior to the seizure.

'You had a short seizure Vincent, nothing too severe, are you alright?' Linda asked feeling apprehensive still, not so much about the seizure itself but more about the bizarre object stuck inside Vincent's head.

'I'm alright, just cold, I should perhaps go to bed,' he replied, oblivious still to what had happened to him.

'You do that, and I will speak to you in the morning, hopefully we'll get in touch with Adam,' she told him casually.

Just after lunch time the next day Linda and Vincent were not just back in Washington but at Dulles airport, ready to board a US Airways flight to San Francisco. There were a number of reasons for the pair taking this journey, for one, Dr Leary's phone number appeared to have been silenced or even made redundant as it wouldn't even go into his voice mail. Then, when Linda tried his office number the receptionist informed her that Dr Leary was out on special leave but didn't elaborate whether it was sickness or something else. At his home number there was no answer, although the phone did ring. Linda considered going to ask the P.D. if by chance he was reported missing but on second thoughts she considered that she had no grounds to make such an enquiry, despite the fact that her womanly intuition was telling her that something was wrong. Finally and equally concerning for her right now, was the foreign object that she discovered in Vincent's head the previous night. By chance Linda had a contact inside the University of California, Berkeley, where they could run a complex 3D scan. Linda's contact was also happy to help even at short notice however it hadn't been easy for her to convince Vincent about the necessity of this trip, nor for her to afford the cost of the flights considering the late booking. Thankfully the journey that followed was event free and before they knew it the pair were in front of the Department of Neuroscience inside UC where professor Nurnberg was waiting. It was coming up to five o'clock and Linda was mindful of that fact.

'Professor,' she said getting out from the cab, 'we can do this tomorrow if it's too much hassle...,'she attempted to explain.

'Ms Wilson, nice to see you again, there is no hassle when it comes to science. And also the lab is free so we can get on with it if your patient is up to it, shall we?' Professor Nurnberg said.

'Sure, allow me to introduce you to Mr Vincent Monroe,' said Linda as she got closer to Vincent.

'Nice meeting you sir, how are you feeling?'

'A little tired and a little confused, but okay! I trust Dr Wilson brought me here for a good reason,' said Vincent with conviction.

'It's nothing to be worried about! This is just a machine that takes many photographs from all around your head and then we can see if there is any abnormality in your brain, it doesn't hurt,' said the professor being somewhat aware of the level of Vincent's comprehension. Nurnberg was only told a little about Vincent's circumstances, but they sounded very intriguing and so he was looking forward to seeing what else he could discover about him. As a scientist, Professor Nurnberg was circumspect and even sceptical, but then throughout his career he came across stuff that scientific methods were not enough to explain it. Also, via new methods of research and the amazing results that had emerged in recent years, the professor concluded that the human brain was far more complex than previously believed and scientists were still in the infant stage of discovery. As for Mr Monroe he had no idea what to expect and after being given a brief description of the abnormality found in his head he could only think of some sort of tumour although he had no real understanding of it.

'So what are we dealing with here,' he asked once inside the lab.

'The swelling is behind the left hand side of Mr Monroe's head, right behind the ear, it's not very visible but you can feel it once you touch it, it is bizarre, it feels as if some sort of object has been inserted there deliberately.' Linda explained.

'I see. Can I have a look before the scan?' asked Professor Nurnberg as he seated Vincent on a chair. He then meticulously pulled latex gloves over his hands and began feeling the indicated area. The professor spent a good few minutes in doing this, checking other areas as well and writing something on a small note book a couple of times. His face expressed nothing but confusion, however he kept quiet all the time.

'Whatever it is, it's very sharp and no, I can't say that I have come across anything like it before,' he concluded, as he eventually broke silence. 'Are you alright Mr Monroe, does it cause you any upset?'

'You can call me Vincent! I don't know if upset would be the right word but it gives me visions sometimes, terrible visions,' said Vincent with aplomb.

'Visions, what sort of visions,' the professor asked genuinely confused.

'Maybe I should explain more about Vincent's circumstances,' Linda interrupted, 'um, he was found in South America by a few locals. I happened to be there doing some research work. I sent Vincent to Washington a few days later. You see Professor, Vincent claims to have been abducted by what we now call extraterrestrials, but not only that, he also says that it happened while on an expedition that reached deep into Peru and it happened about one hundred years ago. I am sorry, but I couldn't tell you that on the phone,' she explained.

'We need to get him to the scan, whatever is there is as strange as the story you just told Dr Wilson,' said professor Nurnberg not quite convinced of what he just heard. Minutes later after he had run through the routine with Vincent and had him settled inside the MRI scanner, Professor Nurnberg and Dr Wilson were watching the first images on three large LED screens inside the monitoring room and so far there was nothing very spectacular, apart from the colourful details offered by 3D imaging. Powerful computers were already processing the data sent by many cameras inside the scanner and slowly, on the screen placed in between the other two, a 3D model of Vincent's brain was taking shape. It started from the base upwards, while the image kept on rotating every few seconds. The other two screens were showing not only images but a fair amount of information about the speed at which the cameras were operating, the subject's vital signs such as his pulse for instance, and a lot more. Every few seconds first hand images of Vincent's brain were sliding across these two screens as the computers were setting each picture in the 3D environment while on the middle screen the shape was becoming more complex with every passing second. Even for such a long time practitioner as Dr Wilson, this was quite amazing stuff and she was now realising that she needed to do some catching up if she ever wanted to get back into working inside a clinic.

'This is something else, this technology, I have never seen anything as accurate as this scan before,' she said by way of speaking her mind.

'Yes, we love what we do here, a few more years of hard work and we'll complete a brain map so complex that we will need to rewrite our neurology text books,' the professor said, his voice enigmatic.

'What are we seeing so far Professor, anything out of the ordinary?'

'Well yeah, I've noticed a few things but I won't comment on anything until the whole process is completed. You see so far the computer is only building the general structure of his brain without focusing on any particular area, but in a few minutes we should start seeing more detailed images from the region of the brain that it is of interest to us. I instructed the computer to do that once it is done with the general view. Perhaps another fifteen minutes or so, we'll see,' he explained. 'Now, do you believe him, about being abducted by ET?'

'I think so, or maybe, but I know where he was found and that was bizarre and then in DC they carbon tested his clothing and the hair found on his clothes containing his DNA, and guess what..., these items were one hundred years old, give or take,' Linda answered.

'If all this is proven this will change the course of history you know?' Professor Nurnberg commented in passing, focusing on the screens. 'We're almost there,' he said a few minutes later. He then quickly began to zoom in on a small area of the occipital lobe, on the left side. It was there that an unusual formation could be seen, something that looked more like an object than anything else. But there was more, it appeared that there were some ramifications going from the strange object way inside the temporal lobe and perhaps even deeper.

'Wow, that's something you don't see every day,' exclaimed the professor his voice somewhere between excitement and confusion. 'Ms Wilson, have you spoken to anyone about him, I mean to the authorities..., this is out of this world.'

'Um, as I said, Vincent is in the care of a state facility in Washington and I believe he is about to appear in front of a commission very soon, but other than that I don't know. What is it that you see professor, what you make of it,' asked Linda, pointing at the screen.

'I don't know, it's nothing that I've seen before and I may require a couple of my colleagues to assist me to read the data, hope that's okay. Now what we see here in the occipital lob on the left is something which I wouldn't be far wrong in saying, is unnatural and perhaps made. How it got there without affecting the subject I can't tell and excuse my ignorance but I don't think anyone could. It is a rather large object that was inserted into his brain and probably Vincent should not be alive. But more intriguing are all these dark features that appear to spread out into the temporal lobe and perhaps all the way into the frontal lobe, well that I could describe as a tumour, not that I've seen one like it and trust me I've seen lots, but that is organic in nature, like living tissue,' the professor explained, maintaining a mysterious tone throughout. He then shook his head and got up, walking from one end of the room to the other, occasionally having a glance to the large screens.

'Ms Wilson,' he eventually said, 'I think what we have here is way out of our league, if this man was indeed abducted by ET's I strongly recommend that you inform the authorities, I don't know, the government, all I can say is that the foreign object in his head could not be surgically implanted by human techniques. We have evolved indeed, in terms of medical procedures and developed new operating technologies that are advanced but not to the level I see here, this is different,' he concluded.

'I need to take him back to DC, they'll take it from there,' said Linda almost at the same time as this awful noise erupted in the reception area, getting closer and closer. It was a mixture of stomping feet and loud voices, all men as far as one could make out. Next thing the door banged open to reveal five men who quickly stepped inside the monitoring room. And as if this wasn't enough Linda nearly fell to the floor when she recognised that amongst the four strangers was Adam Leary who stepped right up next to her. She could not even react as it was like something had pinned her down to the chair, frozen from inside out.

'It's better this way, trust me.' Adam whispered as he came closer.

'Can anybody explain what the hell is going on here, Dean Patel...?' Professor Nurnberg asked, his voice fading away as he saw the last man to step in.

'Surely,' began this tall man in the expensive suit, probably in his mid-30's and seemingly very different, 'you are conducting unauthorised tests on a subject who is not even part of your jurisdiction Professor Nurnberg, I believe.' he said sharply. He inspired nothing but pure coldness, even fear. The other four men were pretty ordinary people, such as Adam Leary, the dean of the Neurology Department at UC, and two security guys, but him, well he did not belong.

'I can really make a mess of this or we can come to a simple arrangement gentlemen, I trust that wisdom will guide you in making the right choices. As for Mr Vincent Monroe, you may take him out of the MRI scanner as soon as I have finished, he is supposed to be in the custody of St Elizabeth's Hospital in Washington and yet here he is in California, perhaps Dr Wilson had something to do with it...?'

'How dare you! Who are you anyway?' Linda asked, disgustedly.

'That is not important, what I can do, is really what you should care about. Anyway my name is Marcus, and that's all you need to know, I am a private investor, high tech and heavy industries with large applications in the medical sector.

Regarding Mr Monroe, well he is of unique importance for some new research and development, and I am in possession of some documents that attest to the fact that he will now be transferred from state care into private care at an undisclosed location. Furthermore, I trust that any results found in the scan that you have just concluded will be transferred over to me and then erased. Dean Patel has given me his assurance that no word will come out of this facility..., ever! This in exchange for a very generous two million dollar donation that Mr Patel was more than happy to accept, is that correct Dean Patel?' Marcus inquired without even turning to face the dean.

'Alright then,' Marcus resumed his speech as soon as the dean nodded at his back, 'Ms Wilson, I will be happy, along with Dr Leary, not to press charges on you for literally kidnapping Mr Monroe, on the contrary I would like to help you continue your tremendous work over in Brazil so there is a cheque for half a million should you wish to accept it...,'

'Piss off you condescending prick,' whispered Linda through her teeth.

'Thank you, I don't need to sign that then. Finally going back to you Professor Nurnberg, I was told that you are one of the best in the field of Neuroscience and Neurosurgery, and Dean Patel is happy to give you leave of absence for an inderminate period of time, should you wish to join my team?

A prolonged moment of silence followed Marcus' speech as if no one dared to say anything or at least nothing that might upset the man in expensive suit. He was intimidating if not actually scary, composed beyond reason, keeping one cool tone of voice throughout.

'Well,' he started in the same manner, 'if you don't mind releasing Mr Monroe I will gladly escort him to the helipad where our transport is awaiting. Mr Nurnberg you have until this time tomorrow to make up your mind or at the least I expect you to sign our confidentiality agreement. Dr Leary, I appreciate your cooperation and rest assured that your patient is in good hands. As for Ms Wilson here she may return to Washington in the next available flight for which I will happily pay and that is all gentlemen! You all have my gratitude for reaching a reasonable agreement in such short time, thank you all!' Marcus concluded turning towards the door.

'Who said we are all in agreement,' asked Linda visibly irritated.

'Ms Wilson you are in the weakest position to raise any objections. Let me remind you that if I am to press charges, you will be going to prison and honestly I wouldn't want to see a smart, beautiful woman and a remarkable doctor wasting years of her life in the lesbian wing of some place in Pennsylvania, that would be such a tremendous loss to our society,' said Marcus sarcastically.

'Give me a break Marcus or whatever your name is, on whose authority are you acting cause so far you've been very vague,' she replied in absolute disgust.

'Huh, let me see, oh yeah, the papers that enable me to take Mr Monroe into private care are signed by no other than Senator Wilson, your ex-husband and the man in charge of the Mental Health Commission of the Senate. Is that to your satisfaction, Ms Wilson?' said Marcus remaining sarcastic still.

'You sanctimonious bastards,' she whispered as she made her way out. 'I can purchase my own ticket,' she shouted over her shoulder. She made a few steps into the reception area and then turned around to face the group of men following her. 'I would appreciate it if I could say "goodbye" to Vincent,' she requested sincerely.

'By all means,' replied Marcus.

Chapter Sixteen

He could not recognise this place, the same jungle but different. This was way too vast not in length but in height, and then there were screeching howls coming from all directions and there were many, horrifying screams of the creatures from hell, if that word would be adequate to describe it. It was dark only for the odd rumble that was preceded by fierce lightening. Only in these moments could he see right up to the top of the strange trees crowns that were shooting up into the sky for tens of meters. Even the sky was unfamiliar as if too close to the stars for they were big, giving a jade aspect to the sky. Then there were many moons, way too close, perhaps on an imminent collision course with one another, and implicitly with whatever this celestial body was, which surely could not be Earth although it was similar. The air was breathable in spite of being heavy with burning smoke, the moist soil under his feet felt like that of Earth, the vegetation was abundant, similar to that of the rainforest in which he once got lost, never to return but to end up here, wherever this place was. For certain this place was the land of hopelessness and pain for there was no way out, none that Vincent could see. He would have rather stayed inside the pit, senseless and without pain, in pure darkness and without vision, rather be dead with no chance of ever coming alive, for living in this moment was beyond painful, it was incongruous, unearthly and showed no light at either end, if there even was an end. This was nothing else but a battleground where Vincent could sense the suffering of many creatures, amongst them the "evil gods", for there was no winner, only a struggle for survival. The rumbling and lightning kept going sporadically in random places, sometimes farther away, sometimes closer and he kept on running, flanked by two of the "evil gods". It reminded him so much of different episodes that he had been part of back in Cuba, but this wasn't the same. Here he was not in charge, had no control over his fate but was left at the mercy of chance. No one on this battleground was in charge they were all running for safety, or at least it appeared that way, as they were all running in one direction. And also one could sense that something had gone terribly wrong in so far as this didn't seem to be part of the plan. There was no strategy and there were no combatants but something beyond all the known forces, seemed to be at work, something far greater than the "evil gods" themselves. If there was any certainty in this place it was the fact that death was coming for them all, no matter from where they had descended, because there were many species as far as Vincent could see, amongst them "the greys", the same creatures he saw climbing into the pits back on the "sky boat". Vincent could not help but take a moment in all this chaos just to observe that out of all the creatures running, "the greys" seemed to find it the most difficult task as their bodies were like those of children, but their heads were oversized, even in adult terms. Their movement was bizarre if not hilarious, but in the midst of all this nothing was amusing. They kept on running as Vincent was pushed and shoved by the two "evil gods", sometimes even dragged across the damp terrain. It all came to a sudden halt at the edge of the forest, in front of this vast plain spreading as far as the eye could see. It was here that one could see the scale of the devastation as lumps of burning rock were falling from the sky, making the ground explode on impact. It was apocalyptic and judging by way the "gods" were rushing, the worst was yet to come. Those moons up above were coming way to close to one another now, suggesting that the collision was just about to happen. There was hope it seemed because several of those "sky boats" were lying at ground level in the valley below.

That was where all these beings were running, as the "sky boats" were the only things that could take off and get out of this hell. However it appeared that not all beings were welcome because up in the air, several fighter orbs were shooting balls of energy continuously at selected targets, pulverising them on the spot. Up near the entrances into the "sky boats" sentinels were also spreading out fire balls, making many approaching creatures just vanish. But it seemed that they wanted Vincent back on the "sky boat", back in the darkness of the pit, as the two beings flanking him were now carrying him, with his feet inches above the ground. It was a better option, pit was good Vincent decided. But suddenly he could no longer see the way in, or anything else except for this strong light exploding right there in front of his eyes, blinding him in an instant.

*

'How is he,' asked Marcus from his comfortable chair inside the command room.

'We have induced the coma, vital signs are looking good, he's good to go,' the young female beside him replied.

'I wish we could see his dreams,' Marcus commented.

'...we're maybe going to see more than just dreams, phase two will commence in twenty minutes. Once we make connection with the alien device, we will see right inside him, deep down into his soul,' the woman said mysteriously.

'Good work everybody, good work,' said Marcus.

'Whatever we're going to see, from the moment we make that connection, will change our lives forever, so be ready for that folks!' Professor Nurnberg said.

'Let's have a break folks, come on all of you, have a break before we commence phase two,' said Marcus.

'Yes good idea, sweet dreams Vincent,' the young female added.

The blinding light was becoming less severe on the eyes now, but still not to the extent that Vincent could make a clear distinction as to whatever it was he was seeing. These were mainly shapes and shadows alternating in front of his eyes in no particular order. He was certainly lying down on his back and, as the vision became a little clearer, he realised that the strong light was coming from something similar to a surgical lamp. He wasn't felling much, just wooziness that was not unpleasant but was not necessarily comforting either. But then anxiety suddenly began to settle over him as soon he realised that two of the "evil gods" were nearby, each of them appearing to check on some instruments that looked of a medical nature. They were very quiet, which made Vincent feel even more apprehensive. Suddenly this illuminated globe on which indecipherable symbols were appearing and disappearing at regular intervals, exploded right above his head, floating in thin air within inches of his eyes. Vincent was scared once more, not knowing what to anticipate, although lately he had experienced nothing ordinary. His biggest problem was trying to determine when all this was happening, because ever since he came across these beings he could not put any order into the events that followed. Was this a moment before going into the pit or after? Or was it before finding himself running to safety amongst all sorts of beings, in a place with a sky full of moons and stars that were all falling on top of one another? Ultimately, was it just his mind lingering on things that took place in the past and he was now re-enacting them in a dream of some sort? It could be, and it would only make sense whether this was, dream or reality, if this moment followed all the others, or else he would not have had any knowledge of these events. The most annoying thing however was the fact that he had absolutely no control over any of these events, ever, and so it was now, so he had no means of fighting or preventing whatever it was that was coming. In the absence of any power over his fate, he just let his mind go back in time, way back to his childhood years. It was then, when every Sunday morning he would go to mass with his mother, and sometimes his father, whenever the old man wasn't busy with military matters. Each time having to listen to stories from the Holy Book, stories about a God who was ever so merciful and loving of Man and whatever the story they told, the Man was wrong and God was right. Furthermore the priest would ask each time what were the teachings of these stories, and there was only one answer, taking many forms, more or less it said that loving one another the way God loves us all, was the way forward. He had grown to know different and found no love in his fellow humans, but now he found no love in the "gods" themselves either, so where would all this leave Man? Pretty much it would leave him nowhere, instead Man would be better off just realising that those stories were poisoning his veins with lies, and furthermore they would leave him in the agony of his own wrong doings, for there is no redemption, there is no love, one for another, and there is no love from the "gods". The world as Vincent had grown to know it, such as it was on Earth and the same in Heaven, was nothing but a violent place, a place of struggle for survival, for it was only the strong who will make it through and live another day. There was no room for love or mercy; there was no God testing Man's devotion to Him, but only room for making Man stronger for the fight which goes on and is never ending. The stage was set a long time ago, the players were all those who had received the "gift" of life, gods amongst them. Who would win the fight? It seemed that the whole purpose of it all was not winning at all but just staying alive, using whatever means necessary to achieve that. So who wrote those stories and for what reason? It could have only been Man, depicting these tremendous events, but his interpretation of these events must have been inaccurate, as he, Man, well he always felt all alone and was desperately in need of someone or something to love him, how wrong, but how could he know any different? It took the kind of experience that Vincent acquired, ever since coming across these beings, to comprehend the meaning, to see the purpose and, as the Holy Book had said, perhaps the one single piece of truth, "Many are called but few are chosen", and those chosen ones will receive illumination, should they choose to accept it? Neither Abraham nor Joseph were tested for devotion or character, but perhaps chosen to learn the truth, a truth that humanity is still refusing to accept for it is more convenient and comforting to build upon concepts such as love and devotion. Vincent learnt only in more recent times that such concepts are foreign to the "game", although he had played the "game" most of his life. And now he was still lying on his back, staring at the energy ball floating above his head, waiting for his fate to be decided upon. The two beings turned their faces towards him and shortly drops of warm fluid began invading his eyes, making his vision go blurred. A very thin needle penetrated the corner of his left eye and went deep inside, releasing something right into his head; something that Vincent could not determine but he felt that whatever it was it began communicating with him. He started to see numbers in front of his eyes, random formations of just zeros and ones but no other numbers. Then he felt this tremendous pain at the back of his left ear and it made Vincent feel very sick, but even more than that he felt that this "thing" inside his head was now growing, taking over his thoughts and even his soul. He didn't feel much more after that as he went into a deep sleep in which only darkness was present and nothing else.

'Connection made, alien artefact is now online, come on tell us what you've got,' the voice of the young female scientist, her name was Lena, sounded in all the speakers around the control room. In an instant all the large monitors inside the room flickered several times as if experiencing some power fluctuation. Then moments later the screens filled up with unknown symbols and then pictures of locations on Earth it seemed, and it went on showing all sorts of things, some looked like mathematics, some looked like blue prints of some kind and some looked like genetic imprints. All in all there were shades of information about lots of things and it was hard to make any sense of them at first glance.

'Power up the auxiliary systems now,' Lena shouted in the background, 'we may experience a "traffic jam" and we don't want that! Oh Holy God, that's a lot,' she commented in amazement.

'First hand material from the "makers", I'll be damned!' Marcus said.

'It's like everything, look at those numbers, I think they are coordinates and dates of landing zones worldwide for God only knows how far back in time,' explained Lena not totally convinced that she had got the right meaning.

'How about the maths in the blue prints, what do you think that's about,' asked Marcus.

'They are more like engineering data of some kind, I can't tell but it looks like machinery to me,' another scientist answered.

'You bet, I hope you're damn right because I choose to believe that they want to teach us how to travel, faster and further into the stars. On the screen left hand side bottom, those seem to be coordinates of different systems in the cosmos, wouldn't you say?' Marcus enquired.

'It would appear so, but I am more curious about the genetics,' said Professor Nurnberg.

'Yes, I was meaning to ask you, since you have had the time to look at that "tumour", what do you think it's made of? I know you said something about being organic, but do you have any more on it?'

'I did do a little more research but cannot yet offer a conclusion. Instead I formulated a theory, so treat it as such until further research is conducted,' the professor replied.

'Go on, try me,' said Marcus.

'Well it is organic but it would appear that it's a complex of metals that I can't determine at this stage. Furthermore I believe that the material is self-aware and it has the ability to replicate itself, if that makes any sense,' explained Nurnberg.

'Can you remove it?'

'I wouldn't dare, I am afraid it will kill Vincent, in fact I have no reservations about that.'

'Perhaps, but our purpose is far greater than Vincent's life, that material is the essence, it's everything, the alpha and omega of all things, so I would say we remove it for further examination,' said Marcus with conviction.

'But sir, what if it does not survive our environment? There is a strong possibility that we need to consider,' the professor explained in consternation.

'I'm sure that you will find a way to contain it without having to expose it to the environment. I also have the feeling that once it senses the danger, it will retract inside the artefact. It originated inside it, didn't it?'

'We assume so but it doesn't mean it will go back,' said the professor.

'As every living thing my friend, we all look for cover when danger is imminent.'

'So how about Vincent, we just let him die,' asked Lena.

'If he dies,' replied Marcus confidently, 'but even so don't you think is worth the sacrifice? For the first time in history they, the gods, found a way to communicate with us at a mathematical level. Why, because it is only now in recent times that we have begun to read these symbols, because we are now ready and sufficiently advanced scientifically to replicate their technology, don't you see? And that "tumour" as Professor Nurnberg calls it, well that's everything, self-aware and self-replicating material, that ladies and gentlemen is The Source. We've chased it for centuries and now I've got a lot more than I expected. I thought that they, the gods, were The Source but no, they only had access to it and now they've sent it to us so we can share its' magic, so that we can make another giant leap into the future,' he explained with much aplomb.

'Now you sound like those lunatics from the History Channel I'm afraid.' Lena commented.

'Huh! Them? Well they have never got near anything that we've got but I am sure they'd die for this sort of evidence.'

'Look Marcus, you did it, you got tons of information downloaded from your "Source", but having to kill Vincent is just immoral,' said the professor.

'Oh no shit? Why don't you try to put a list together of whatever humanity has done that was of high moral value because I can tell you it's going to be a very short list. The world did not advance based on moral actions!' Marcus shouted, reducing the professor to silence.

'It will maybe take days before all that info can be downloaded. How is our subject, is he gonna make it through? I am afraid that if we disconnect him we'll compromise the data,' said Lena.

'Intravenous feeding every five hours, he should be okay, he is a tough bastard that Vincent fellow. If we decide to move ahead with surgery I suggest that we should never "wake" him up,' the professor explained.

'I agree,' Marcus added sharply, 'and as long as we have sufficient space to store the data we should not be worried about processing....'

'Huh, that will take many years, and with these computers I just can't tell,' one of the young male scientists said.

'If you'd let me finish young man, I was saying, processing will be done at the IBM headquarters by a quantum computer, it will be a lot shorter than you think!' Marcus stated seriously.

'Wow, I never thought that was ready, they....,'

'They always have things ready before they are even officially announced as being in development.' Marcus added.

As the darkness began to fade away once more, for poor Vincent none of that talk was of any concern. He was connected to the quantum reality of his own experiences as his subconscious was jumping from vision to vision in no particular order it seemed, but just giving him a hell of a time while masses of information was being transferred to several data storage servers from his alien implant. It was raining abundantly now and it felt more familiar whatever this jungle was that he found himself in now. He was in pain, severe physical pain as if he had been run down by a train. Once more he had no means to determine the location or the time frame but somehow he felt he no longer belonged here, not in this jungle. It appeared that he had left it a long time ago when he saw the end of it and then began climbing onto the rather bare hills and running across bushy plains. But somehow he was now back, back deep inside the rainforest, the same rainforest that claimed the life of many of his comrades and almost claimed his. Suddenly this question was becoming more obsessive with every minute that passed; did he ever leave this jungle? The last time he saw it was not long after spending some time with a local tribe, during which he and his mates went through some tough times, experiencing hallucinations. What if it all had been nothing else but hallucinations? Perhaps now the locals had just abandoned him here in this place, all alone it seemed. He had to entertain the idea to some degree, simply because he could not explain how he got here, but something was telling him that what he experienced was far beyond hallucinations. He must have lived through events that were hard to imagine and almost impossible to explain. He saw things that were way beyond the understanding of this world, he met beings that were not of human origin and suffered horrors that would have certainly left him scarred beyond healing. Some of these horrors he desperately tried to deny and to some extent he did such a good job that now these were fading away, stuck inside the episodic buffer of his memory, as if he had witnessed someone else going through these horrors. He eventually got cold as the rain became heavier and he thought that he had better found shelter rather than keep trying to determine his location in space and time. It was certainly a vision of Earth, so finding shelter should be an easy task because he was very familiar with this terrain. He struggled as it seemed that he had sustained many injuries which were hurting, but he eventually found shelter. He knew from that moment that it was only by chance that he would stay alive because in the state that he was and in this environment, death was imminent. As for chance, it was perhaps too much to hope for. Nevertheless he was still here now, wherever this was. And perhaps this "here now" was beyond the realm of relativity.

Something like an inner voice, was telling him that this wasn't the real state of affairs but just visions from a past time, hard to establish. After this moment the only clear recollection he had was that of human faces, strangely decorated, but certainly human. They were talking in a bizarre tongue, poking at him with sticks while he was desperately trying to hold on to a backpack. He knew that whatever the bag contained was of value, like some sort of testimony of his experiences hidden inside of it. Once more the faces and the voices faded away only to be replaced by other voices, speaking his native tongue this time, but he had no image of those whom the voices belonged to. Some sort of a metallic noise was accompanying the voices, which was getting louder and louder with each moment that passed. He wanted to scream as he sensed that he was in great danger, perhaps the moment in which he was to die, this time with no hope of return, this time with no room left for visions, this time he would breathe his last breath, that of sorrow and regret for not being able to explain in his own terms what it was that had happened to him. However he didn't want to feel sorry, dying was a good option, but dying in total confusion was dismaying. Moreover he knew now that he wasn't just dying, but he was being killed, and he was being killed by his fellow humans without warning, without reason it seemed, and without compassion. It was part of the "game", part of the same "game" that all those who received the gift of life became "players", as on Earth and so in Heaven, he was just another "player". A lightning ball struck him so hard from the top of his head right down to his toes, making every part of him shiver violently. Flashes of memory started to roll out in front of his eyes, people, gods, bizarre creatures, one after another, and each time he felt like he was travelling at high speed, away from them all, as if he was going backwards while they were standing still in one spot. But it all came to a sudden end, his vision becoming stuck on one strange character; a cowboy in an immaculate white suit, wearing a white hat with a large brim and holding two silver pistols, both pointed at Vincent.

'I got you now, didn't I? Welcome to my world old pal, I got things to show you, most unpleasant things, but you need to die first, so die!' the evil character said, releasing one bullet from each gun, both hitting Vincent right in the chest.

TO BE CONTINUED